Human Resources Management in Construction

9

Human Resources Management in Construction

D. Langford, M. R. Hancock,
R. Fellows & A. W. Gale

Longman
Scientific &
Technical

Longman Scientific & Technical
Longman Group Limited
Longman House, Burnt Mill, Harlow
Essex CM20 2JE, England
and Associated Companies throughout the world

First published 1995

British Library Cataloguing in Publication Data
A catalogue entry for this title is available from the British Library.

ISBN 0–582–09033–4

Set by 5 in 10/12 Baskerville
Printed in Malaysia

Contents

The authors and acknowledgements

The authors would like to thank the following for permission to reproduce copyright material: Institute of Manpower Studies for fig. 3.1 from 'Emerging UK Work Pattern' by Atkinson J. (1984) in *Flexible Manning – The Way Ahead*, IMS Report No. 88; Sage Publications, Inc. for fig. 3.2 from *Culture's Consequences: International Differences in Work-related Values* (abridged version) edited by Hofstede G. (1980); Scientific Methods, Inc. for fig. 4.4 from the Leadership Grid Figure in *Leadership Dilemmas – Grid® Solutions* by Robert R. Blake & Anne Adams McCanse (1991) (formerly the Managerial Grid figure by Robert R. Blake & Jane S. Mouton) published by Gulf Publishing Company, Houston; F. E. Fiedler for fig. 4.5 from *A Theory of Leadership Effectiveness* by F. E. Fiedler (1967) published by McGraw Hill, New York; Prentice-Hall, Inc. for fig. 4.6 from *Management of Organizational Behaviour: Utilizing Human Resources* (4th Edition) by Hersey P. & Blanchard K. (1982); American Society of Civil Engineers for fig. 4.7 from 'Motivating Construction Workers' by Laufer A. & Jenkins G. D. (1982) in *Journal of the Construction Division, ASCE*, Vol. 108; Gower Publishers for fig. 7.3 from *Handbook of Management Development* by Mumford A. (1979); MCB University Press for fig. 7.4 from *Management Development: Theory and Practice*, Bradford Management Bibliographies and Reviews by Ashton D., Easterby-Smith M. & Irvine C. (1975).

The authors would like to acknowledge the work of Victoria Langford for an intellectual contribution to Chapter 7, F. Goudie for critical comments and discussion, and Bernadette Cairns and Kristeen Clayton for their work in preparing parts of the manuscript. Thanks are also due to those academic and industrial collaborators who have assisted in the development of research projects from which many ideas and arguments have originated.

D. Langford

Professor David Langford has 17 years' experience of teaching and researching in construction management and is actively engaged in management development for construction industry professionals. He holds the Barr Chair of Construction in the Department of Civil Engineering at the University of Strathclyde. He has written extensively in the field and has co-authored several books such as *Construction Management in Practice* (1982) and *Strategic Management in Construction* (1991). He has presented papers on the subject at venues around the world and has presented many short courses to senior construction professionals in the UK, Asia, Australia, the USA and the Caribbean.

M. R. Hancock

Dr Mick Hancock has 20 years' experience in the construction industry, including 10 years in higher education. He held posts as a quantity surveyor in both the public and private sectors before entering education working in both the UK and Asia. He is currently teaching, researching and consulting in the area of construction management and building economics. His particular interests include the self-determined development of Third World construction industries and philosophical aspects of the building process and its management. He is currently a Lecturer in the Construction Study Unit and Director of Studies for the MSc in Construction Management at the University of Bath.

R. Fellows

Dr Richard Fellows has worked in several contracting organizations and has 17 years' experience researching and lecturing in construction management. He is currently Senior Lecturer in the Construction Study Unit and Director of Studies for the MSc in Construction Management by distance learning at the University of Bath. He has published a variety of books on construction management and contracts, and has presented papers, particularly in the field of construction economics, at numerous international conferences. His current research interests include pricing of projects, effects of culture on project performance and marketing.

A. W. Gale

Andrew Gale has 13 years' industrial experience in Britain and the Middle East working on civil engineering and building projects in project management and cost engineering roles. He has worked as an academic since 1985 and is currently a Lecturer in the Department of Building Engineering at the Manchester University Institute of Science and Technology. He teaches at undergraduate and postgraduate level as well as devising and running construction management training internationally. He has carried out research into women in construction, Russian construction management and building quality management, publishing and presenting papers in Australia, Sweden, Israel, Norway and at other venues.

Chapter 1

Introduction

The diversity of schools of management and the social sciences share a common base of concern for people and their behaviour – both individually and collectively. Due to the dynamics of societies, not only are birth, death and taxes certainties but, perhaps, the most important certainty is 'change'. Change generates uncertainty and risk about the future; often, data are sparse and so decisions are frequently highly judgmental. Managers are concerned about the future as the future can be influenced, if not changed radically; the past and the present are fixed. So, managers make decisions, necessitated by and effecting changes, which necessarily concern people.

Construction is a labour-intensive industry. The degree of labour intensity varies from sector to sector, project to project, country to country but usually within quite narrow boundaries. The concept of labour intensity is relative between industries. Common measures are value of output per person; value of output per operative; value of plant and equipment per operative; output as percentage of GDP relative to percentage working population in the industry. Such measures are not without statistical difficulties – definitions of the boundaries of the industry; obtaining data from the many very small firms. Many consultants consider themselves (and are regarded and 'counted') as associated with but not part of construction. Clients, increasingly, employ construction personnel in-house; many large private sector and numerous public sector organizations have a construction division or department; many individuals undertake DIY activities and most countries' construction industries have a 'black economy' at work.

Hence, people are the foci of alternative views of the industry. Construction exists to contribute to the satisfaction of human needs and wants; it is organized by people; it employs people. Only relatively recently has the pervading importance of people begun to be recognized in essence and extent. It is the personal interactions which generate demands and determine the nature of supply responses.

Whether humans can, or should, ever be regarded as resources is complex and questionable. This book considers a wide variety of construction-based issues of human interactions and examines how such interactions can be managed.

Human groups

It is all too easy to regard an organization, or, indeed, any group of people, as having its own identity, distinct from but incorporating the identities, goals and values of the individuals of which it is comprised. Whilst many groups do have a corporate identity, corporate activities remain subject to influences from individuals and sub-groups. An overview of construction, as in Figure 1.1, provides classification of human groups as Clients (public/private; large/small; individuals/corporate), Consultants (architects; engineers; surveyors) and Constructors (building/civil engineering; main contractors; subcontractors; management contractors/design and build contractors; operatives/managers, etc.). Clearly, other groups are involved less directly – financiers, insurers, planning authorities, tenants – although the influences of such groups are being recognized increasingly.

Clients

For all but very small projects, the client is unlikely to be an individual. As Cherns and Bryant (1984) noted, corporate clients are complex organizations which include many sub-groups. Once a project reaches the industry (i.e. emerges from the client as a design brief), it has undergone much scrutiny and debate. It is likely that the project represents a 'victory' for the sponsoring group within the client organization and, hence, a defeat for other groups who, due to capital rationing, may have seen their projects put into abeyance or abandoned.

Within the client organization there are vested interests some of which will seek to accentuate the success of the project and others the converse (to the advantage of their project proposals in the future).

Broadly, clients are either experienced and/or expert (sophisticated) or naïve. Expert clients build often and, commonly, employ construction professionals in-house, know what performance can be demanded and how to obtain the required performance; such clients 'drive' projects. However, the majority of clients are naïve, they build very infrequently, know little of the industry (except via image) and may be influenced easily – by advertising or/and advice of the first contact with construction personnel (usually an architect). Hence, for naïve clients, those who obtain first contact with them regarding a project are in a significantly advantageous position.

Consultants

Normally, the consultants' group comprises those concerned with design, primarily architects, structural/civil engineers, services engineers, quantity surveyors – but the group will expand to incorporate project managers, planners (town and programming) and various other specialist service providers. Most projects require inputs of various design specialisms whether they be provided through consultancies or by a design and build contractor.

In project chronology, the design phase shows the first major manifestation of functional fragmentation typifying the construction industry. The interface problems between separate design organizations are subject to the over-riding factor of the major interface with the client – usually the responsibility of the 'lead consultant' (architect in building; civil engineer in civil engineering). In a study of architectural practice, Mackinder and Marvin (1982) found that most clients (the majority being naïve) do not know what information should be provided to constitute a (good) brief

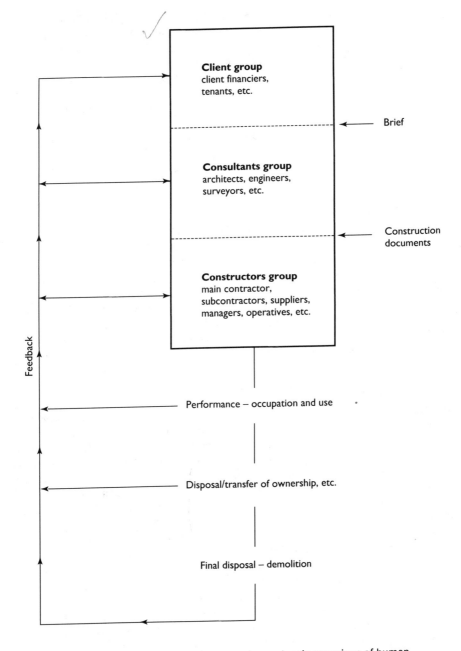

Fig 1.1 Overview of the provision of construction and main groupings of human resources

but, much more alarming, most architects do not know what briefing information is necessary.

General tendencies

Initial decisions tend to have the widest ranging effects: Kelly (1982) noted that about 80 per cent of the cost of a building was committed by around 20 per cent of elapsed design time. Hence, attention to early decision making is of great importance. Perhaps too commonly, with little external guidance and often in a matter of a few hours, the architect determines the arrangement of spaces and other preliminary design aspects using, primarily, aesthetic criteria.

In seeking to overcome design problems, Mackinder and Marvin (1982) found that architects seek solutions in the following sequence:

- own experience
- immediate colleagues' experience
- product leaflets
- practice library
- research findings and own research

Clearly, the heavy reliance placed on experience does raise the question of what constitutes experience – at least, what is remembered. Recall may be quite selective, especially in the context of incidental (experiential) learning, as is likely to occur in work situations. Condemnation, perhaps due to the vehemence of its administration and associated guilt feelings, is likely to be remembered much more strongly and clearly than praise; further, many managers condemn much more readily and frequently than they praise. Hence, designing from experience is likely to have a very much stronger input of avoiding repetition of failures than of repeating successes – a conservative approach, tending to perpetuate the 'safe' *status quo*.

General consequences

There has been a tendency amongst consultants to carry out functions sequentially; the required inputs from various consultants occur discretely and so, coupled with obtaining such objectives as planning permission, etc, the design phase becomes protracted. Project finalization activities (agreeing final account, etc., which involves the quantity surveyor primarily) may be 'stockpiled'. Hence, in various ways, consultants are both able to cope with unexpected urgent tasks and to have work to fall back on during slack periods. Such approaches facilitate constant employment of resources in the consultant firms themselves but the design phase is elongated notably over what could be achieved by more parallel working, as in NEDO (1983). A significant proportion of contractors' funds are therefore tied up for unnecessarily (and hence costly) long periods – a notable proportion of such funds is written off each by contractors.

It is apparent that the desire for full and continuous employment of their resources amongst consultants (an imperative under fee bidding and during recession) has consequences detrimental to clients and to constructors: to clients through protracted design periods and conservative design tendencies; to constructors through extended periods of financial 'lock-up'. Unsurprisingly, there are benefits too: longer design periods afford greater opportunity for iterative design development and evolution (testing alternatives, etc.), fully employed resources yield a version of high productivity which should reduce prices (fee bids) and delays to (final) financial settlements on projects and aid clients' cash flows on individual projects, although, to remain viable businesses, contractors are likely to seek recovery of such shortfalls on future jobs!

General constraints

Frequently, consultants (particularly architects) seek to be and regard themselves as being aloof and separate from the construction industry. Seeing themselves as consultant (artistic) designers, they are cast in the role of guardians of cultural creative input to the built environment, charged to tell clients what they really want (what is good for them and the community) and determining what constructors build. Cultural custodians to the exclusion of carbuncles!

Increasing involvement of architects (and other design consultants) in the expanding sphere of design and construct serves to temper extreme eccentricities and to foster appreciation of others' requirements and the consequent enhancement of designs achieved through that approach.

Constructors

The organization of construction has been transformed over recent years: fragmentation of activities has become markedly more pronounced to the extent that the main (general) contractor provides the primary construction management for execution of construction work by subcontractors, even for primary trades such as bricklaying and carpentry.

Current tendencies

Since the late 1970s, employment in construction operations has become more casual and fragmented, reversing the trend of decasualization of employment (which had persisted since the 'navigator' era of the 19th century); a trend which has been embodied in employment legislation, such as the Employment Act, 1980. A casualty of the return to fragmented, more casual employment has been operative training; de-skilling remains a highly contentious issue despite the greatly reduced requirements for many traditional skills (e.g. thatching) and the needs for new/different skills (e.g. dry lining).

Consequent upon the structural/process changes which have occurred amongst constructors, is a shift in the industrial power-base. Due to their metamorphosis into management contractors, the general contractors have, to a significant degree, relinquished their control of project price determination (if it ever was within their real control); such power now rests in major subcontractors and suppliers. In the long period, in a capitalist economy, all businesses must earn normal profit as the minimum return on investment for survival. However, in the short period – such as forms consideration of an individual construction project – much more pricing freedom exists in order to bid keenly, to buy work, to 'take cover', etc. Thus, for an individual project, the main contractor retains the position of arbiter of the price to be bid (and the primary consequence-bearer thereof).

General tendencies

The traditional model of price determination shows price to the predicted cost plus mark-up, i.e. estimate plus adjustments plus 'allowances' plus profit. However, particularly in extreme environments (recessions or booms), market conditions are the primary determinants, within limits, of the prices for individual projects. Greater variability still pertains to final prices of projects which, to a significant extent, are determined by the negotiating skills of the people involved and the relative (market) power of the organizations.

Government

Clearly, the operation of the industry is highly dependent upon the interactions between the human groups which are involved with the project directly. However, a variety of other groups are influential, if only by exerting major influences upon the project environment. In such a context, the government is the most important: it acts as direct client, as indirect client and as the primary influencor of the business/ social environment.

Impact of privatization

Although government's involvement as a client, both direct (e.g. Local Authority Housing; road construction) and indirect (e.g. Hospitals, via the National Health Service), has diminished in the UK largely due to privatization as a mainstay of policy, in the early 1990s, the public sector was client for around 35 per cent by value of the industry's work. Through monetary policy (notably interest rates), employment legislation and further implementations of socio-economic policy, the government remains highly influential over the 'health' of the industry.

Recognition of the impact of government has been manifested in the contributions to Conservative Party funds by major construction organizations (a notable amount of which decreased as the economy spiralled into recession with construction's taking an extensive downturn in workload, prices and employment). Unlike many other industrial groups (e.g. Agriculture), the construction 'parliamentary lobby' is not strong and so, it appears, that withdrawal of contributions to the political party publicized to be the stalwart of private sector market enterprise is the most effective way of protesting and attempting to secure more favourable policies.

Impact on trade unions and employment

Due to the success of 'anti-collectivism' since the late 1970s, and the structural and technological changes within the industry, the position of trade unions has been diminished. Growth in sub-contracting and self-employment has placed main contractors, and similar employing organizations in a stronger bargaining position (manifested in 'pay when paid' clauses, etc.). Such enhanced power (and consequent profit-potential) is subject to countering influence through the increased power of clients due to their increasing knowledge, sophistication and expertise (to demand and secure enhanced project performance in terms of completion dates, higher quality, lower prices), which in turn are magnified by the consequences of economic slumps.

The overall result has been an industry subject to even more liquidations and bankruptcies than usual, including large organizations (e.g. Rush and Tompkins). Further, takeover activity has assumed a new dimension – that of construction companies from other parts of Europe taking major stakes in UK-based contractors. UK construction organizations, traditionally the domain of domestic capitalists and, often, with large interests abroad, have become less insulated from international take-overs with the consequent loss of control and subjugation of UK-based interests to those of the overseas owner (rather like an imperialist-reversal).

Relationships and industrial processes

As people are involved so extensively in all facets of construction, it is important to examine how people, whether as individuals or grouped in organizations, relate to each other and to the processes and procedures adopted in the industry. For such an examination, behaviour, value structures, desires and practical constraints are important. As people change, sometimes both rapidly and extensively, much is transient. Whilst such considerations are vital at the level of the individual, due to the effects of aggregation, they should be less important for groups (especially large ones). Hence, the social sciences express generalities of behavioural traits, value structures, etc. with assurance.

However, assumptions about things such as behavioural goals (e.g. maximizing satisfaction) may conflict and, hence, require a decision over which is to be adopted or involve reconciliation. (The outcome of such a decision may well be influenced by the view of the decision-maker.) The resultant behavioural profile will form the basis for deciding future courses of action.

Planning

Decision-making is problematic. This is because decisions are made by and about people, involve the future, are concerned with change and use imperfect information and knowledge. As decisions are invariably based upon prediction, it must be recognized that predictions, by their very nature, contain inherent errors no matter how sophisticated the technique used to produce the prediction. The aim should be to obtain the predictions likely to have the least error to support the decision, given the constraints of time, cost, etc.

Unfortunately, the construction industry has a (largely justified) image of unso-phisticated decision making. At one extreme, an individual makes a decision on 'experience' or 'gut feeling'. Commonly, construction decision-makers use forecasts derived from extrapolation of trends coupled with allowances for known forthcoming events (e.g. a statement in the Budget) plus built-in experience of the decision-maker. The industry has not only been quite slow in the adoption of computers but also in using them to the full for forecasting and risk management techniques.

The increasing expertise and demands of major corporate clients are expressed through expanded knowledge of the industry and ability to penetrate the image of the industry and individual organizations. Clients are the driving force. Many adapt procedures to suit their own requirements (e.g. with tendering: interviews of contractors' project management teams, following bidding, to select the most suitable contractor) or develop their own procedures/documents.

Centralized decision-making

The complexity of the multitude of relationships, contractual and otherwise, which have existed on many projects with the consequent obscuring, transferring and 'sharing' of responsibilities (making the task of seeking and obtaining remedies highly problematic) has led to a widespread desire for a single point responsibility. Undue desire for simplicity of resource channels and use of standard procedures which do not take adequate account of clients' current requirements have precipitated introduction of new procedures and processes (design-and-build procurement routes, construction management, etc.). A particular development amongst clients, which demonstrates more sophisticated analysis in awarding projects, is for sums of money to be 'assigned' to non-price considerations (e.g. good performance on previous projects or, on the other hand, a high incidence of submitting contractual claims). Such 'soft money' is used to adjust the contractors' bids to determine the predicted overall prices which are considered for selecting the contractor.

Quality

A factor which has been subject to much attention during the late 1980s and early 1990s is quality, whether as 'Total Quality Management' (TQM), or, more frequently, 'Quality Assurance' (QA). The usual manifestation of QA, consequent upon BS 5750 and EEC provisions of ISO 9000, is for organizations to develop procedures to ensure that they will provide, as quality, what is specified. The volume of the QA procedures' documentation is, all too commonly, regarded as indicative of provision of good quality due to the existence in the organization of an authenticated QA system. Unfortunately, such systems provide no assurance that appropriate quality is specified! TQM seeks to overcome the quality issues more globally by considering what quality should be specified, that it is specified and, then, achieved: here a QA system will be very helpful, if followed.

Management tools

The usual way of providing a required level of output quality has been to check items progressively as production takes place. Items of inadequate quality are then either scrapped or reworked. But in the view of Deming (1986), provision of quality output by removal of (most) defective products is, merely, treating the symptoms; the essence of achieving quality is to get things right first time, thereby avoiding waste through scrappings, reworks and much of the need for progressive and continuous quality checks. The imperative is to change the approach of the people – to get them more involved (through 'quality circles', etc) – so that they have the desire and incentive to produce quality. Usually, the approach requires abolition of 'payments by results' schemes (i.e. piecework pay). It is the role of management to foster the involvement of operatives and, by providing facilities and a conducive environment, to secure improving and high quality output. An important element is a forum for operatives to suggest improvements (to products and processes), for them to be considered and acted on appropriately so that, by knowing they are inexorably involved, all personnel have an enhanced commitment to the organization and its output. Trust and permanence are vital to securing such commitment and the consequent improvements/advantages.

The 'Just in Time' (JIT) production approach requires ensuring that inputs arrive at the workplace only when required, the aim being to reduce holding stocks of components (inventories). For JIT to operate, a smooth and reliable provision by suppliers is vital; hence, a high level of confidence in suppliers' abilities to perform is required. Rigorous scrutiny of suppliers is employed to ensure that the system will not fail through suppliers' defaults. Naturally, where a production process is subject to frequent and significant changes (such as on a construction project site), the potential for using JIT is restricted severely. However, with on-site processes assembling components which have been manufactured off site, the potential for using JIT can be high with consequent overall gains for the industry as a whole.

Risks

Clearly, major changes are continuing to occur in the construction industry. Risks and uncertainties are being re-assessed and redistributed amongst the parties involved. Despite some attention to Risk Management techniques since the late 1980s, there is precious little evidence of their formal adoption by project participants. Risk management which is employed occurs naturally (as do many other management techniques) through managers' experience/expertise/intuition. Even then the formalities which are used occasionally by a few organizations comprise only the first stage of identification of risks, sometimes extend to the second stage of quantifications of risks but the stages of allocation and response are left to the contract provisions and price determination processes.

Risks should be identified, quantified and 'allocated' to the contractor only if the contractor (or subcontractors) is able to control those events. Such 'allocations' act as incentives for good performance. A further requirement for such allocations is that the contractor is able to bear (perhaps through insurance) the consequences if the risk actually happens. Uncertainties, however, (being events, the possible occurrence of which cannot be predicted by using statistics) cannot provide performance incentives and so should be borne by the client. People have differing degrees of aversion to risk; an individual's level of risk aversion can vary with changing circumstances. Hence, risk management involves combining objective measures of the risk faced with the decision-maker's perceived values (positive as well as negative) of those risks.

As clients become progressively more expert, their use of more sophisticated, 'scientific' management methods or tools increases. Risk management, life cycle costing, decision theory, etc. are becoming more widely employed. Clients want accurate predictions. Unfortunately, the very nature of forecasting, the variability and individuality of projects and the nature of the data-bases available in construction mean that, virtually always, forecasts of construction project performances contain large degrees of variability. Although recognition of such variabilities (which are greatest for forecasts made early in the project, i.e. during design) is present in forecasts being provided in the form of the 'most likely performance' with a 'tolerance range', the most usual form for performance forecasts remains 'single figure' predictions. For safety, the single figure predictions include large contingencies whilst giving a false impression of accuracy. Two of the potential consequences are that the client is enticed to make over-provision for the project or that the presence of the safety net of contingencies encourages project performance to require those contingencies to be used, thereby fostering a poorer level of performance than could be obtained.

Procurement

Commonly, procurement methods have been regarded as ways of organizing projects with necessary incorporation of specific forms of contract, methods of determining the price and of paying the contractor. However, with alternative payment systems, price setting mechanisms, the proliferation of different forms of contract (both 'standard' and *ad hoc*) and the wider variety and awareness of methods for organizing the project, it is appropriate to regard procurement methods as a selection of sub-systems. The result is greater flexibility, individuality and suitability but with the requirement to evaluate alternatives and decide upon the most appropriate. Expert systems are being developed to assist in the selection of procurement methods.

Figure 1.2 shows the use of main procurement systems whilst figure 1.3 shows use of major standard forms of building contract. Although there is no necessary relationship, the norm continues to be for particular contract forms to be used for certain procurement methods. The forecasts for use of the procurement methods were obtained through questionnaires, interviews and futures research with clients and contractors. The noted decline in the use of the traditional procurement method (design-tender-build, normally using JCT80 or similar contracts) is forecast to continue. Use of management methods (construction management and management contracting) will remain approximately constant whilst the notable increase is in the use of design-and-build methods (design-and-build; design-and-manage; design, manage, construct, etc). Results given are measured by value of projects.

Whether the trend away from traditional procurement is a consequence or a cause of increased organizational fragmentation remains an issue of debate. Despite 'anti-lump' legislation (legislation against tax evasion by self-employed subcontractors), the 1970s onwards witnessed increasing fragmentation amongst constructors through the reaction of main contractors to increased costs of direct employment (occasioned by employee protection legislation) and to cost-increasing risks caused by large and rapid fluctuations in workload (influenced by governments' stop-go policies). Notably, a countervailing trend is apparent in design consultants with the formation of multi-disciplinary practices and with the emergence of design-and-build constructors.

Although quite different, the strategies represent 'niching' with the previously traditionally-based main contractors focusing on their project management expertise whilst the multi-disciplinary design practices and design and building constructors focusing on increased degrees of single-point responsibility (from the client's view) for project provision.

Since the mid 1980s, contract awards have been based on more comprehensive evaluations of probable performance than those epitomized under tendering codes,

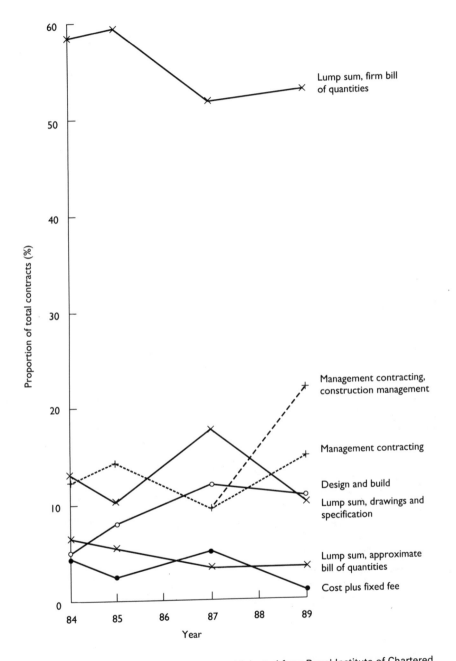

Fig 1.2 Use of main procurement systems. (*Adapted from* Royal Institute of Chartered Surveyors' statistics)

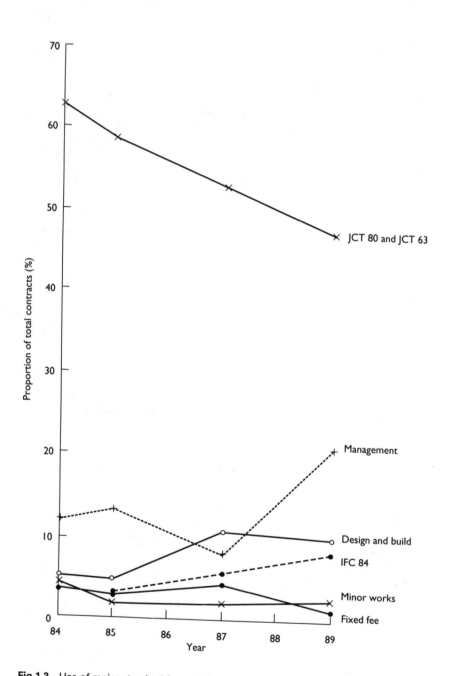

Fig 1.3 Use of major standard forms of building contract. (*Adapted from* Royal Institute of Chartered Surveyors' statistics)

etc. (notably *Code of Procedure for Single Stage Selective Tendering*), whether for selection of constructors or of consultants. The interviewing of/presentations by intended project teams assists clients in evaluating non-price factors which, in conjunction with price elements, are used in the final selection of which organization(s) to employ.

Marketing

The abolition of fee scales amongst consultants and the reduced role of price in selection of constructors has exerted pressure on the industry to move from selling to marketing; a logical and appropriate extension of which would be into Relationship Marketing (see, for example, Gronroos (1991), Fellows and Langford (1993). However, it is evident (Lim (1990), Fellows and Langford (1993)) that the industry remains highly selling oriented, regarding promotion and selling as synonymous with marketing. Baker and Orsaah (1985), Fellows and Langford (1993) found significant discrepancies between what clients want from constructors and what constructors believe clients want. Following Ohmae (1988), for example, market success requires producers to orient themselves to fulfilling customers' needs (demands) to the greatest degree possible. If producers have distorted or inaccurate views of what customers want, meeting demand and achieving customer satisfaction is a random process!

Normally, marketing has been oriented to production industry although some attention has been devoted to the particularities of market services. Especially with the changes in the construction industry, the marketing of services is apposite.

It has been common for marketing to take a systems-wide approach, requiring analysis of the supply environment and of the particularities of the item/organization constituting the supply. The approach has required analysis of the environment by considering the PEST factors (Political, Economic, Social and Technical; legal factors should be included). The factors relating to the supplier are subjected to SWOT analysis (Strengths and Weaknesses of the supplier – internal factors; Opportunities and Threats in the market(s) to which the supplier should respond – external factors).

The relevance of supply to marketing

Usually, analyses have concentrated on 'technical' aspects – time of delivery, quality and price, epitomized in the 4P's of marketing – Product, Price, Promotion and Place. Such aspects constitute the functional performance of an item. For the majority of larger construction projects, especially if located in or/and executed by designers and constructors from developed countries, differences in instrumental (functional) performance between suppliers is likely to be marginal/negligible. Hence, the major constituent of client satisfaction, as argued by the relationship marketing school, is the method of supply, the primary element of which is the relationships between the client's personnel and the supplier's personnel – or the expressive performance. Generally, it seems that the more involved are client's personnel with a project, the greater is their satisfaction with the level of project performance achieved, that is personal involvement enhances a person's perception of the performance achieved. So, given adequate instrumental performance, it is

13

good expressive performance which is vital and conducive to the forging of good and continuing business relationships; and particularly so as suppliers are very keen to have regular customers who have repeat orders and so provide a steady base level of demand!

International activities

Improvements in transport and communications have fostered international trade. For many years, UK has 'exported' construction by both designers and constructors executing projects overseas. The increasing internationalization of construction in Europe has been spurred by provisions of membership of the EEC. UK imports construction materials and goods as well as construction services, directly, such as in the presence of Japanese constructors, and indirectly, through take-overs of UK constructors by companies based overseas.

Even for the largest construction organizations, as with the major Japanese conglomerates, there remains considerable doubt about whether a construction organization can be 'global'. (Indeed the requirements for an organization to be credited 'global' appear to be flexible and it may be that no organization really is global!)

Certainly, a major consequence of the changes is for enhanced performance and flexibility. There is much greater awareness of what performances can be achieved (although the mechanisms of town planning, and the consequences of safety measures, are not always understood) with the result of more demanding clients and considerable proliferation of (the components of) procurement routes.

Structure of this book

Following this chapter of an introductory overview of certain primary issues concerning human resources management in construction, the remaining nine chapters consider contextural factors, managerial issues, and recent and future developments. The topics explained and discussed employ findings from research to supplement the presentations of theoretical constructs and evaluations of techniques and procedures used in the industry.

Contextural factors (chapters 2, 3 & 4)

This section of the book analyses how human resources management relates to the construction industry and organizations within and related to the industry. Appreciation of opportunities for action and the constraints which apply is vital; knowledge of the sources of and reasons for the opportunities and constraints provides basic contextural awareness and, hence, where, why and how any changes may be brought about. Chapter 3 focuses on the wider, external issues affecting demand for construction and supply of construction with emphasis on the people-based processes involved. Interactions of people with technology are considered in the context of work requirements, patterns and employment protection within a capitalist environment. In Chapter 4, attention is on organizations – their structures, designs and operations in a variety of construction industry contexts. Organizational behaviour is determined by the people comprising the organizations and the powers of human groups. Of particular importance for organizational success (however determined) are the relationships between people – involving leadership, followership, motivation – the internal formal and informal structures and procedures of the organization.

Managerial issues (chapters 5, 6 & 7)

The section discussing managerial issues concerns policies and processes regarding the human resources of organizations. Chapter 5 traces the framework of the industrial relations machinery within the construction industry and maps the changing contours of the industrial relations scene. This paves the way for company-specific activities designed to promote good relationships between employers and employees. Chapter 6 examines the process of recruitment – from determining requirements within organizations, through obtaining applications, sifting the applications, selecting the most appropriate people. How they are selected and developed is the subject of Chapter 7: management development in the industry. The chapter includes determination of who are the most appropriate people to participate in programmes of management development, what the programmes should aim to achieve and the mechanisms available plus how the success of such development programmes may be evaluated.

Recent and future developments (chapters 8, 9 & 10)

This section notes what changes have occurred in human resources in construction during recent years and what seem to be probable developments in the future – both extensions of existing changes and novelties. Chapter 8 examines developments occasioned by 'new technologies', notably those of data handling and processing and of communications. Although technologies applied to construction processes continue to evolve at increasingly rapid rates, research (e.g. Fellows, 1993) suggests that robotics will remain of limited use with primary developments in that field being centred on Japan. A notable area for development and application of robotics in construction is to cope with hazardous operations, such as those involved in de-commissioning nuclear plants. Not only will CAD become ever more widely employed by both designers and constructors but expert systems will be developed for commercial use (usually for diagnostic purposes) and clients will adopt more 'scientific', technology-based decision-support processes, including risk management, value management and decision analysis.

In Chapter 9, the developing roles of women in the industry are examined. Numerous barriers remain to women entering the male-dominated, macho-imaged industry. However, the potential contributions of women as design consultants, managers in construction organizations as well as operatives is receiving wide recognition and, albeit grudging, gradual acceptance. The development may well present the industry with one of its greatest challenges and richest sources of changes despite the highly adverse image and lack of knowledge of the industry presenting major barriers to recruitment (of men as well as of women).

Chapter 10 looks to the future of how people will shape and work within the construction industry. A perspective of a variety of human/technology interfaces and interactions is presented with a view to analysing appropriateness of alternative managerial styles.

Summary

To introduce the subject of human resources management in construction, this chapter has provided a contextural overview of the environment; the industry; peoples' roles, activities and relationships. That organizations are made up of people and that people provide the active factor as customers, suppliers, decision-takers and bearers of risks is fundamental. Due to its labour intensive nature, such primary considerations are magnified in construction. Individuals' perceptions and behaviour have been discussed in a variety of settings, demonstrating their impacts on levels of satisfaction, marketing, relationships and project performance. The structure of this book has been sketched to demonstrate the logical flow of the topics covered and the pervading, but all too often subjugated, importance of people – the human resources.

Questions

1. Discuss the main components of procurement systems and examine the factors which modern, sophisticated clients are likely to consider when selecting a main contractor for a project.
2. Evaluate the importance of an organization's personnel in its marketing activities.
3. Explain and discuss the contention that reliance on experience promotes conservatism in design.

Bibliography

Baker, M. and Orsaah, S. (1985) How do the Customers choose a Contractor? *Building*, 31 May

Building EDC–NEDO (1984) *Faster Building for Industry*, HMSO

Cherns, A. B. and Bryant, D. T. (1984) Studying the client's role in construction management, *Construction Management and Economics*, **2**, 177–184

Deming, W. E. (1986) *Out of the Crisis: Quality, Productivity and Competitive Position*, Cambridge University Press

Fellows, R. F. (1992) Contractual changes in construction in Cannon J. (ed) *Construction to 1996*, NEDO

Fellows, R. F. and Langford, D. A. (1993) *Marketing and the Construction Client*, Chartered Institute of Building

Gronroos, C. (1991) The marketing strategy continuum: towards a marketing concept for the 1990s, *Management Decision*, **29, No. 1**, 7–13

Kelly, J. (1982) Value analysis in early building design in Brandon P. S. (ed), *Building Cost Techniques: New Directions*, E & F. N. Spon, pp 115–125

Lim, P. L. H. (1990) *An investigation into the 'Qualifiers' and 'Winners' of Construction Projects*, MSc Dissertation (unpublished), Construction Study Unit, University of Bath

Mackinder, M. and Marvin, H. (1982) *Design Decision Making in Architectural Practice*, Institute of Advanced Architectural Studies, Research Paper 19, University of York

Ohmae, K. (1988) Getting Back to Strategy, *Harvard Business Review*, Nov–Dec

Chapter 2

Human resources management in the context of construction

In the 100 years or so that management has been a subject of study many different and varied aspects have been thought about, written about and implemented. Despite all the advances made, no single theory of management exists and it is common for writers and managers to view the individual strands of management thought in isolation. A division of subject areas is, of course, necessary in order to attain a clear view of certain particularities; indeed central to the whole concept of management, is the ability to break down work activities into manageable and identifiable parts. However, another important aspect of the management process is the subsequent integration of those divided parts in order to achieve better productivity from the labour process. Managers must therefore view each of the various subdivisions of their discipline in the context within which it operates.

This book is concerned with the case of human resources management (HRM) in the context of the construction industry. As such we will emphasize the nature of HRM in terms of the organizations within the industry and the environment within which that industry must function.

The relationship between human resources management and the external environment

The construction industry is responsible for the production of the built environment. As such it designs, manufactures, maintains and demolishes all the man-made buildings, bridges, roads, tunnels, dams, etc. that we utilize every day of our lives. Without humans there would be no built environment of the kind that we see today. In all built environments there is some kind of order and planning, usually of a kind most suited to the needs and situation of the community for which it was designed.

At every stage of the building process there is a need for (at least some) planning, organizing, forecasting, control and co-ordination. These, of course, are the original functions of management as propounded by Henri Fayol. Whatever arguments have taken place since Fayol's work was published (particularly concerning the area of managerial roles), it is still true that these activities are required for successful management. The important thing to remember is that this is not all that is required of the modern manager and to rely only on such vague terms is to invite problems at all stages of any kind of production. It is in recognition of this point that many writers since Fayol have considered approaches to management that consider the organization of work as more than a straightforward mechanistic process. Today's writers allow for the unexpected, or in other words for contingencies that may arise at the work place.

Capital and human resources: fundamental differences for management

Whilst the management of capital equipment may be complex, it does tend to be essentially consistent and dependable. The greatest difficulties arise when we try to manage human beings. Humans do not behave in a regimented or mechanistic manner and therefore create a number of difficulties that are not apparent in the management of plant and machinery. The major resource input to construction is human and it is therefore important for us to understand the peculiarities of this entity in terms of how it affects the managing of the building process.

If humans do not behave in a completely predictable manner, then how might we describe their activities? Human behaviour is governed by a unique kind of mental programming which arises from a combination of three broad sources of influence. We can refer to these influences as:

- general
- genetic
- group influences

General influences are those that affect all humans alike and which are therefore common to the *Homo sapiens* species. Genetic influences are those that we inherit through breeding and from our forebears. There is a degree of commonality in these two spheres of influence, but perhaps the most significant area of mental programming is that which derives from group influences. Because we all have different life experiences we will tend to view ourselves and our world differently (however slight this may be). The behaviour of humans is, then, individually conditioned by their history, surroundings, schooling, upbringing, financial position, social grouping, breeding, the law, etc. So we can say that humans react to a variety of stimuli that induce certain kinds of behaviour and that these stimuli are the product of environment.

The systems view

Whilst humans are affected by their environment, so their behaviour and activities have an influence on that environment, thus maintaining a continually changing and interactive relationship between the two.

The theory used to describe this idea is 'systems theory' and an understanding of the process of interaction between subject and environment has led to changes in the way that managers in all fields approach their task. Increasingly managers see projects and organizations as input/output systems; inputs coming from the resources used and outputs being completed buildings and/or successful organizations. In taking this approach they recognize the effects of the environment on both the inputs and outputs and vice-versa (Figure 2.1).

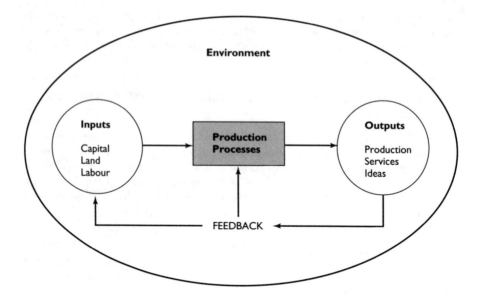

Fig 2.1 Relationship of inputs to outputs

In understanding the notion of systems, managers appreciate the complexity of the real world and are able to identify possible intrusions to the system (from the external environment) over which they may have no direct control.

Survival for the firm depends on its ability to interact with this external environment. Briefly, we can identify four general facets of environment:

Economic: interest rates, inflation rates, unemployment level, competing firms for market share.
Social: attitudes and values of employees towards work, attitudes and values of clients towards business and products, employees' level of education and training.
Political: laws and regulations governing the firm's operations (including employment practices, wages, etc.), industrial relations, health and safety.
Technical: availability of raw materials, plant and machinery, market demands for particular manufactured (secondary) material inputs.

Whilst interaction with the environment is essential for survival, a firm's chances of success depend more on the ability of management to achieve an optimum degree of fit between the complex and sometimes conflicting organizational objectives and culture, individual employees' abilities and expectations, type of work and external environment.

The complexity that systems theory allows us to map is one that shows itself very clearly in the construction industry. Figure 2.2 shows how the external environment within which a project or individual site is situated, becomes part of the internal environment when we look at the building firm. The building firm is situated within a separate external environment, which in turn becomes part of the internal environment when we prepare a systems

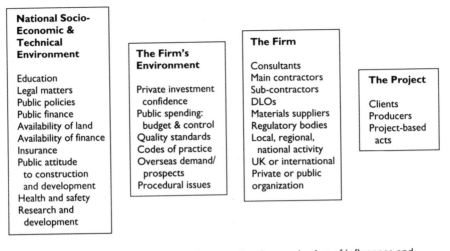

Fig 2.2 Overall systems view of construction – showing a selection of influences and variables. Note that the number of variables demonstrates that building is rather less in control of its environment than most other industries

model of the construction industry as a whole. What soon becomes apparent is that:

- The problems facing managers are complex.
- Systems are recurring by nature.
- At whatever level a manager is operating they must maintain stability between the internal (human) and external (influencing) environments.

Effects of 'turbulence' in the environment

The construction industry is prone to economic fluctuations just like other industries. However, much construction work is labour-intensive, so booms and slumps in the economy tend to have particularly severe effects. Not one single building firm can claim to be in complete control of its external environment and therefore its destiny. Whilst the industry as a whole has devised methods of indicating likely business cycles, usually based on historical economic data, these only give the broadest of outlines regarding possible movements in the market for building and in interest rates, etc.

Since the Oil Crisis of 1973 the environment within which construction-related industries exist has undergone changes that have probably been more rapid and disruptive than at any other time in recent history. A major influence in this has been the very close relationship between the industry and the economy. In times of economic change, variations in construction demand and output affect levels of employment and wages (and vice-versa). These factors are themselves related to other economic issues, such as the balance of payments and governmental policies concerning capital expenditure programmes. Here we can see the inter-related nature of what we might term macro and micro environments. At one level governmental intervention (or lack of it) and the balance of payments (macro environment) will establish a set of conditions that will be faced by all participants in construction. Although these conditions are the same for all, their responses will vary according to the specific circumstances and situations faced by the individual organization or project. This then is the micro environment and consists of interactions between clients, consultants, suppliers, sub-contractors, employers and employees, etc.

Newcombe *et al.* (1990) describe environments as being polar in nature, ranging from stable to turbulent. They consider the former to be a rarity where with higher (than usual) levels of certainty about the future, changes 'tend to be evolutionary rather than revolutionary.'

More common, and certainly the case in the last twenty years, has been the existence of a turbulent environment. The rapid, disruptive and seemingly unpredictable nature of these changes have required construction firms to become more flexible, responsive and innovative. This has not gone unnoticed at an industry-wide level. In 1976 and 1978 two studies undertaken for the Building and Civil Engineering EDCs (*How Flexible is Construction?* and *Construction in the Early 1980s*) were concerned with identifying the implications for construction-related industries of 'possible levels and patterns of demand in the early 1980s' and to understand and analyse 'the resources required for different types and stages of work, the extent to which resources can respond flexibly to changes in demand, the capacity of the industries, time lags, etc.' (*How Flexible is Construction?*).

Certainly changes in the operating policies of construction firms can be seen in the increasing use of subcontracting and the development of new and varied procurement systems, but these changes have tended, on the whole, to be reactive rather than proactive in their nature.

Environmental scanning

In an industry where human resources are one of the major inputs, an understanding of how the external environment is likely to affect this key input, and the impact it will have on the firm's effectiveness and profitability, is critical to survival and success. One means devised for the purpose of increasing insights into the character and outcome of changes in the external environment is 'environmental scanning'.

The Harvard Business School has, for some time, considered environmental scanning to represent the starting point in describing a planned pattern of objectives and policies that shape a firm's present activity towards future superiority in competition. In short an environmental scan comprises the identification and analysis of current environmental trends in order to evaluate their influence on HRM in the future. Much of this book is in fact a kind of environmental scan, in which current trends and aspects of HRM will be identified that we believe will have a significant effect on the success or otherwise of construction firms in the next decade.

Combined environmental trends: identifying and analysing

Although certain individual changes in the external environment (e.g. government policy, public opinion) can lead to direct consequences for the management of human resources, it is much more likely that changes are brought about as a result of combined trends. It is not unusual for trends emanating from differing areas to not only affect each other, but even to be co-determined. As with systems theory, it is necessary to take a holistic view of the remarkably interconnected environment in order to get a meaningful picture that equips a firm to meet the environmental challenges of the future. Scanning at this level is likely to include the monitoring of publications and broadcast media, attendance at conferences and even the maintenance of a wide range of contacts in the industry and associated fields.

Typical of the areas that a firm would scan are socio-economic, political, legal, demographic and educational trends. Each of these will be dealt with later in the book, but for now we can illustrate the kind of things that we might look for under some of these headings.

Demographic changes, for instance, will affect the size of the available workforce as changes occur in birth and death rates, life expectancy, migration, etc. Linked to this we might see changing trends in education that will alert us to an over- or under-supply of qualified labour in the future and might then persuade firms to take a more active role in the training of its labour force. These two areas may well be linked to each other and changes in both of them may be the result of political policies. All three of these may then account for alterations in discernible socio-economic patterns. The point here is to reinforce the ideas of interdependence and complexity mentioned earlier.

The examples above constitute what we might call macro-environmental trends. For successful HRM, firms will also have to consider the micro-environmental aspects. Among the areas for consideration at this level might be questions about the firm's competitive position in the industry; its long term planning strategy; its managerial philosophy; the changing nature of technology and any noticeable trends in the behaviour or attitude of its employees. For example a firm might construct what Porter (1985) refers to as an 'industry scenario' in order to appraise its competitive environment and thenceforth to design its future competitive strategy. By scanning its internal environment, employee attitudes may come to light that indicate the need for changes. For instance, an apparent lack of morale and commitment on the part of employees may be a manifestation of a perceived reduction in the quality of their working lives. Alerted to the problem the firm will be in a position to rectify the situation, perhaps through a restructuring of its operations and the forming of small groups which will tend to be more cohesive and self-supporting. Regardless of the means used to resolve the problem, the key factor here is in the identification of problems before they become untenable.

General implications for construction firms

Knowing what to scan is one thing, but how often should a firm undertake the process? Should they be scanning on an irregular, regular or continuous basis? This is a difficult question to answer as it depends on a number of factors, each of which may affect different firms in different ways according to their individual circumstances, size and environment. Certainly continuous scanning will ensure the most up-to-date information over a wide range of environmental factors, but the process will be costly and may even require the establishment of a specialist department within the organization.

In the construction industry only a handful of firms are large enough to warrant continuous scanning. The majority of firms are in the small to medium size range and would therefore be more likely to consider environmental scanning on either an irregular or regular basis. Because of its importance to the national economy and its association with virtually all human activity, building constitutes a dynamic industry. As such it is influenced by a wide range of environmental factors and therefore requires more frequent environmental scanning than firms in less dynamic industries. For this reason it is wise to undertake the process of scanning on a regular basis thereby ensuring that trends are perceived early and responses initiated in good time.

As with all data collected by firms, environmental scans are of little use unless their implications are analysed and integrated with the organization's overall objectives. Analysis involves identifying the human resource implications of environmental events and the setting up of human resource objectives. Critical at this juncture

	Scanning period		
	Irregular	**Regular (say annually)**	**Continuous**
Likely useage	Small firms	Medium sized firms	Large firms
Scanning medium	Ad-hoc and occasional studies	Data from occasionally updated historical studies	Structured data collection and processing system
Scanning range	Individual events	Limited events	Broad range of environmental systems
Reason for scanning	Crisis resolution	Decision making	Part of overall planning process
Action category	Reactive	Proactive	Proactive
Time span for effects of decisions made	Current/short term	Short term	Short/long term
Personnel involved	Various staff	Various staff	Environmental scanning staff

Fig 2.3 Environmental scanning matrix

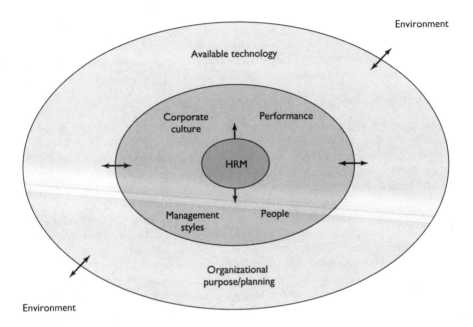

Fig 2.4 Systems view of HRM in the organizational context

is the reconciliation of projected labour demand (in terms of numbers and skills) and likely labour supply. The supply side of the equation is in turn likely to be affected by motivating factors and job satisfaction.

In major firms, the link between scanning and corporate planning may occur in a formal manner. Issues identified during the scanning and analysis processes may be introduced at the beginning of the strategic planning process in the form of assumptions and guide-lines. As planning proceeds, human resources managers will be called upon to check the feasibility and desirability (from a personnel viewpoint) of the plans. Once the corporate plan is agreed the human resources managers will determine the human resource implications and then begin planning for action.

Because only a few firms in construction are of a size to employ specific human resource managers the majority need to integrate their scans into the overall corporate strategy in a less formal manner with managers/owners of firms undertaking both procedures. Nevertheless, the idea behind integration is relevant to all, regardless of their size.

Human resources management within the organization

Along with an appreciation of human resources management in the context of the environment, we need to have an understanding in the context of the organizations within which it takes place. In systems terminology this is the system within the environmental supra-system concerned (in broad terms) with the combination of organizational structure, planning and action necessary for the successful performance of the organization.

This, however, gives us only a vague and basic description. To understand better, we need to look at and ask questions about the various elements within an organization that affect the role undertaken, policies implemented and activities of the human resources manager.

What is the purpose of the organization?

In construction as in many other industries there are at least two sorts of formal organization that cannot be classified as voluntary or charitable in their nature. There are those which operate in what we refer to as the Private sector and those in the Public sector.

All organizations are established for a purpose. In the Private sector the purpose is usually to satisfy effective demand; that is to undertake work for which they will be paid by clients who are personally or corporately bearing the financial risks of the project. Here the financial success or otherwise of the organization will depend upon satisfying those clients.

In the Public sector the emphasis is on the fulfilment of the needs of individuals or the public at large as identified and funded by government of one kind or another. In this sector profits and losses are usually the benefit or responsibility of the government's coffers.

Although, as Brech (1975) points out, making money is not the prime objective of an enterprise the fundamental difference between the satisfaction of needs and effective demand of the two different organizational types does mean that each will have different aims and objectives, which will be marked in their approach to HRM. While this may be so, in recent times the centrality of economic measures as indicators of success has been challenged and, whilst important, these are seen today in terms of their position *vis à vis* other (more social) organizational objectives.

Human resources managers must understand what the organization's goals are if they are to attempt a meaningful integration of corporate and individual desires.

What is the size of the organization?

The relative emphasis on HRM in an organization usually revolves around this question. In general terms, the smaller the organization the lower the emphasis. A problem here is in defining small, medium and large. Sikula (1976) suggests that small organizations would have less than 200 employees and large ones over 1000. In the construction industry the vast majority of firms are, by this definition, small. As a result the HRM function tends to be undertaken on a part-time or informal basis by line managers within the firm. Medium sized firms are likely to employ some kind of personnel officer in an attempt to provide the function in a more organized way, but only amongst the very large building organizations are human resources likely to be the concern of a separate and specific department.

What is the structure of the organization?

An example of the influence of the external environment on the human element of an organization can be seen in terms of the design or structure of the organization.

Typically, firms in a dynamic industry such as construction tend to have a looser structure than those operating in more stable industries. This is another reflection of the need (mentioned earlier) to respond quickly to changes in the economy, clients' expectations and the marketplace.

Although a fairly loose structure is common in construction firms, organizations still vary considerably in their hierarchical structure and span of control. Some might be described as tall structures with line managers overseeing a small group of employees, whilst others are flat with a large number of employees reporting directly to the line manager. Organizations can also be structured according to geographical area, product or function. On building sites the most common structure is the Project Team structure consisting of members from different functional areas within the body of their parent organization(s). The choice of structure normally reflects the degree of formality, level of supervision and extent of centralized decision-making in the organization. This in turn affects communication and general management in the organization.

What is the role of technology in the organization?

Technology, the means by which raw materials are transformed into finished products and services, is a key factor in determining how the work of the organization is carried out and therefore what kind of people will be employed and what training facilities provided. A number of management theorists, notably Trist & Bamforth of the Tavistock Institute and Joan Woodward, have considered the effects of technology on the organization and have concluded that it has a definite influence on both the culture and structure of an organization. Construction is frequently referred to as a labour-intensive industry, the implication being that there is or has been very little change in the technology used. This is a misrepresentation. Whilst the site-based activities do tend to be more of a laborious component assembly function this is certainly not the case in other areas of the industry. The components used (bricks, precast concrete units, preformed window and door frames, roofing tiles, etc.) are increasingly the result of high technology production. At Head Office the finance, administration, estimating and planning functions are becoming ever more computerized as is the work in the professional consultancies such as the production of bills of quantities, specification and even design. As Fryer (1990) points out: 'Technology has made the human factor more important, not less so.'

Changing technology requires human resource managers to give advice and assistance to the organization in relation to the redesign of job specifications, recruitment and training of the personnel to undertake that work and in coping with those employees whose jobs have been overtaken and rendered obsolete by the changes.

What is the philosophy underpinning the organization's culture?

What is important to the organization in terms of the behaviour of its employees? All organizations will have different ideas, values and beliefs about this which will be demonstrated in the amount of freedom accorded to individuals and in the types of personalities attracted to the organization. The organization's culture will also affect its structure and will have been moulded by the history, nature of business and environmental conditions that have prevailed during its existence. Only through an understanding of the particular organizational culture will the human resources manager be able to successfully perform his/her function.

Indeed, the very relationship between HRM and the organization will, to a great extent, depend on the philosophy and beliefs of the senior management regarding the importance of human aspects of the organization.

What effects will the employees and their jobs have?

Jobs are created by the organization and constitute the activities and tasks identified as necessary for the achievement of the organization's function. Jobs are undertaken by the employees of the organization.

Employees are the people hired specifically for the purpose of carrying out the jobs identified as necessary by the organization. Employees and their jobs are the very *raison d'etre* of HRM. For success, the organization needs to be able to create and maintain some kind of order between a variety of key factors related to these intrinsically interconnected elements.

Central to the performance of employees in their jobs will be the level of satisfaction that they gain from the work they do. Satisfaction will depend on the subjective view of an individual, but broadly we might identify three main factors of influence:

- independence or freedom from direct supervision or control
- challenge or difficulty offered by the task
- diversity or heterogeneity of the work

How each individual employee reacts and responds to the varying levels of each of these job factors will depend on their own unique combination of personal characteristics. There are two main considerations here.

1. The employee's suitability for a particular job in terms of capability, education, training, dexterity, competence, etc.
2. The employee's preferred type of work and employer, based on his/her personal goals, ambitions, priorities, inclinations and character.

Given such a wide range of variables it is not difficult to appreciate the intricacies and extent of the human resources manager's work. Involvement begins from the earliest decision to establish an organization and might be classified as consisting of job-centred and employee-centred activities. Job-centred activities include the identification and creation of the jobs to be done and the engagement of the necessary personnel to achieve the organization's objectives. Employee-centred activities include training, performance monitoring and reward structuring, health and safety programmes, industrial relations negotiating, morale and motivation issues, etc.

The success or failure of an organization will depend to a great extent on the ability of the human resources manager to understand and control the multifarious problems and situations likely to arise from the interaction of the various job performance and employee characteristic components listed above throughout the life-cycle of the organization.

What is the life-cycle of an organization and how does it affect human resources management?

The life-cycle of a human being can be described as the process of being born, growing up, maturing, declining and dying. We can identify a similar process in terms of organizations. First the organization must be created (born). Once alive it must develop and grow. This period of dynamism is characterized by expanding workforce, increasing market share, diversity of products and a growing return or profit for the stakeholders. The third stage, maturity, is a more settled period of consolidation. Here a market niche has been found, the size of the organization is relatively stable and concern is directed more towards maintaining the current position of the organization. Transition to the stages of decline and eventual death depend upon the organization's attention, particularly during the mature stage, to its health. Are sales and profits falling? Is the turnover of labour too high? Has the organization kept abreast of changes in the business and social environments? Are current products and markets relevant?

Just as the organization and its personnel change over time, so the efforts and plans of human resource managers must be directed to ensure the continued well-being of the firm. During the period of birth, concentration will be on recruiting the management personnel who will ease the organization into the market and who have the imagination and drive and flair to push the organization through the difficult early periods into the mature stage. As the organization grows, so it needs to ensure that it employs the best possible operatives. Training opportunities will help to attract the required people to the firm and reward incentives linked to the firm's growth are likely to help ensure loyalty to the organization. The mature stage requires a different, less dynamic approach. The organization is likely to be more structured and the selection of key personnel somewhat less important. As this stage development of both people and products is necessary to sustain employees' motivation, morale and thereby organizational performance. Attention to these issues will be vital in order to forestall the period of decline. If, or when, the organization finds itself in decline there are only two options left open to the human resources manager. Retraining of personnel, alongside a new strategic plan, might save the organization. If this doesn't happen, only enhanced rewards will be likely to keep the workforce and this may, in any case, prove to be self-defeating for the firm if no new plan for recovery is produced.

Who should do human resources management?

All firms employ and need to manage people. Essentially the choice as to who should be responsible for the management of human resources can be reduced to two main sets of options. Firstly, should internal employees of the company or external consultants be used? Secondly, should the work be carried out by an individual or by a team? As the next section of this chapter shows, the construction industry consists of a few large and many small firms. The chances of a small firm finding the direct employment of a specialist cost-effective are remote. The most likely option for these firms will be the allotment of time to the managing director or perhaps (if not too small) the strategic planning team.

Certainly, for human resources management to be successful it must be integrated into the strategic management of the firm. Chandler (1962) found that a major factor in the success of large firms was the creation of full-time strategic managers divorced from operational decisions. In many respects this can be applied to human resources managers whose concern is with the efficient operation of the organization through the use of effective employees. As such they need to concentrate on the functions and activities of the firm and the range of external (to the firm) factors that influence its operations, such as economic forces, labour supply, the law etc., and not on the operations themselves.

Whether HRM should be carried out by an individual or a team also depends to some extent (but not entirely) on the size of the firm. There are advantages and disadvantages associated with both approaches. Individuals tend to make decisions quickly, but are less likely to have considered all the implications of those decisions. On the other hand, a team effort is likely to be better in qualitative terms, but slower in the making and may in fact be a compromise in which no one is truly happy.

Whether internal employees or external consultants should be used also presents us with benefits and disbenefits. External consultants are likely to have specialist experience, have an impartial, fresh view of the firm and be more likely to suggest innovative ideas. Alternatively, they are likely to have limited knowledge of the organization, be seen as intruders and therefore be unpopular with those on the inside and have a limited commitment to the solutions they propose as they will not be required to implement them.

It is therefore difficult to generalize about who should undertake the role of human resources manager, but perhaps not so difficult to suggest a number of attributes that someone carrying out the role should possess. The American Management Association's 48th Annual Human Resources Conference (New York City, March 1977) considered precisely this issue. Although there is no indication of the specific industries represented at the conference a number of the characteristics mentioned by some of the delegates do seem as relevant for construction as for any other industry. Their overall view seemed to be that the human resources specialist should exhibit the following qualities:

- Be an effective manager.
- Be well informed about all functions of the business and able to contribute to the development of strategies aimed at achieving the objectives of the organization.
- Be a good problem-solver and planner, able to make decisions and implement them.
- Have the ability to sell his/her ideas to the senior management of the firm.
- Have an understanding of all levels of the organization and the ability to communicate with each of them.
- Have a flexibility of mind capable of adapting to the changing environment whilst retaining an independence of mind and a well-developed system of personal values.

Whoever fulfils this role, regardless of the size of the firm, will then need to have two apparently diverse sets of abilities: an understanding of legal, statistical and procedural matters and an interest in people, problem-solving and the actioning of decisions.

Constraints for human resources management arising from organizational culture and traditions
---

The British construction industry is large in terms of both output and employment, currently producing approximately 7.5 per cent of the Gross Domestic Product and employing in the region of 1.5 million people (approx 6 per cent of total workforce). It might, therefore, be reasonable to expect it, as one of the nation's largest industries, to lead (or at least to keep up with) developments in all aspects of management. However, HRM has traditionally been undervalued and therefore underdeveloped in the construction industry. The constraints responsible for this situation can be identified at two levels; the macro or industry level and the micro or firm and project level.

Constraining characteristics of the construction industry

The construction industry is a large industry comprising many small firms. Figures for 1991 show that of approximately 210,000 firms recorded, some 98 per cent (205,000 firms) employed less than 25 people. Only 47 firms (0.02 per cent) employed more than 1200 people.

Furthermore, the small firms were responsible for only 38 per cent of the total work, by value, whilst the large firms undertook 15 per cent. This outline (which as Table 2.1 shows, is not atypical) provides a number of clues to the problem.

Cost of human resources management

Firstly, the small firms predominant in the industry cannot afford a formal, separate human resources function within the firm; the cost being prohibitive in relation to pay-roll and other operating costs. For these firms such a specialization is seen as an overhead that cannot be borne, particularly in such a competitive industry and during periods of economic recession. In any case, the man-hour cost of HRM to these firms is insignificant and can easily be absorbed within the general administration costs or be shown as an office overhead. There is simply no incentive to introduce the function.

Even within the medium and large firms there is concern over the costs of introducing a formal HRM policy/department. Central to their worries is the balance between time and efficiency. Traditionally, line managers have carried out various aspects of the HRM function in an informal manner. This has a cost in that it detracts from their prescribed job and lowers their efficiency as they have to spend time accumulating the skills and knowledge required to perform this extra work. Only when this cost is higher than the cost of introducing a formal approach will a firm consider introducing HRM as a specific function. If a department is to be formed for this purpose, establishment costs will immediately show up in respect of office rent, furniture and salaries, etc. This will in turn reflect a direct cost in overheads, a loss in profit and a perceived lessening of competitiveness. These cost implications, then, act as a constraint which will only be surmounted when growth, turnover and overheads reach a level where the cost of forming the department represents a minor but justifiable proportion of overall costs.

Table 2.1 Value of work done by private contractors

Size of firm (employees)	Value of work done (£ million, to 3rd quarter each year)										
	1981*	1982	1983	1984	1985	1986	1987	1988	1989	1990	1991
1	156.4	202.6	310.6	346.6	382.8	454.3	527.8	577.0	700.2	784.5	809.3
2–3	267.9	341.3	431.0	498.2	564.7	628.8	698.4	756.2	918.2	955.0	883.5
4–7	358.9	451.6	567.5	601.2	584.5	663.6	744.3	785.2	814.0	783.0	754.5
8–13	344.4	359.5	386.8	411.4	405.2	433.7	486.9	464.7	477.1	467.5	435.8
14–24	401.3	433.7	504.8	544.5	522.4	563.4	617.9	653.0	727.2	696.6	615.8
25–34	220.8	240.9	272.7	286.1	269.3	287.6	342.7	394.0	421.8	421.7	365.6
35–59	358.0	392.5	431.2	446.4	450.9	520.0	581.8	609.4	658.9	632.8	556.2
60–79	170.1	196.5	233.8	224.7	212.2	241.8	280.2	319.4	350.1	377.3	354.4
80–114	207.5	233.6	239.1	256.6	272.4	282.3	324.7	385.4	516.7	508.7	472.3
115–299	577.5	564.4	600.7	641.3	654.8	698.9	851.3	957.8	1,146.8	1,178.5	974.8
300–599	440.4	414.1	428.7	416.9	462.1	470.8	576.7	647.3	743.3	811.6	835.6
600–1,199	357.0	364.6	371.9	407.4	413.5	478.2	592.2	742.5	827.4	857.8	810.8
1,200 and over	620.2	704.6	675.8	771.3	779.4	803.0	918.8	1,206.0	1,532.0	1,812.8	1,369.2
All firms	**4,480.5**	**4,900.1**	**5,454.5**	**5,852.7**	**5,974.3**	**6,526.5**	**7,542.9**	**8,498.0**	**9,833.8**	**10,287.9**	**9,237.6**

(Source: The Housing & Construction Statistics 1981–1991)

* Information based on the definition of the construction industry as given in the 1968 Standard Industrial Classification. Subsequent years based on the Revised 1980 Standard Industrial Classification.

Fragmented nature of the industry

Large firms may well have a human resources management department at head office. From here the company's overall policy will be developed, but a distinguishing feature of the construction industry is the site- or project-based nature of its operations. Effectively the firm is subdivided into smaller decentralized units wherein individual site managers exercise considerable autonomy. Such autonomy is generally necessary for the achievement of performance targets and is valued by those who hold it as a symbol of his/her success and status. These individuals will strongly resist what they may consider to be undue interference from head office including the introduction of personnel policies that are perceived as reducing their authority. These managers will argue for the convenience of recruiting workers at the site, citing the need to meet production requirements as their rationale. Having selected and recruited the employees, and being closely aware of their individual performance and local conditions, it is natural for the site manager to consider themself best placed to judge and decide upon appropriate rewards or sanctions.

Many large firms are financially structured in such a way that individual project sites are held as virtually self-contained cost/profit centres. Here, autonomy (again) coupled with the relatively small size of each operating unit will lead to the HRM function being undertaken on a day-to-day basis by the site manager.

Labour mobility

Market forces ensure that labour will be able to secure better pay and conditions in areas where labour is scarce. This combined with the temporary nature of project-based employment has led to a traditional mobility of labour in the construction industry with the workforce (including site managers) being prepared or forced to move to another location or firm upon completion of a project. Large firms will have project sites at various geographical locations which may or may not be suffering from a shortage of labour at any given time. This variation in supply and demand makes many aspects of a centralized human resources policy difficult to implement.

Shallow management structure

In her studies of the relationship between technology and management structure, Joan Woodward found that one-off or small batch production industries generally have a shallower management structure than those industries that make greater use of technology through mass or process production. The latter industries were shown to consist of many levels of management to implement a greater number of administrative procedures.

At the project/site level, construction demonstrates the 'one-off' nature to which Woodward referred. Building projects generally require a high level of co-ordination, specialist input and consequently direct communication and high levels of informality between the participants in the process. This necessitates a shallow management structure with few hierarchical levels. This structure (also common in small firms) allows a relationship to develop between the site management and workers which enables discussion not only of work-related issues, but also of social and personal matters. From these relationships the manager can gauge the general feelings of the employees and problems and grievances can be settled quickly and amicably. Whilst this may be the case, managers may overestimate the strength of such personal contact with the workforce, particularly in terms of their authority to resolve matters beyond his/her remit such as disputes over wage rates, paid holidays, training provision, etc. and so the informality should be kept in perspective and not used as an argument against human resource management.

Subcontracting and casual employment practices

As a defence against uncertainty of workload, economic slump and unstable demand, an increasing amount of building work is being undertaken by subcontractors. The days when contractors directly employed the various tradesmen required to perform all but specialist tasks have been superseded by the employment of labour-only subcontractors employed as and when required.

As these firms and individuals are not in the direct employ of main contractors there is no responsibility for them over and above ensuring that they are paid for work done. Indeed this system further fragments the industry, creating more small firms and encouraging increasing levels of self-employment and casual labour. In 1981 the Department of Employment estimated that 24 per cent of labour in the construction industry was self-employed, by 1991 this figure was 42 per cent.

We can see from this that construction firms are able to increase their turnover through growth and expansion without a proportional increase in directly employed personnel. In fact, they are able to achieve this growth precisely due to a reduction in the number of employees. Any additional staff will tend to be in the form of administrators required to handle the increased number of subcontractors. With fewer employees to consider, firms will tend to view the expense of human resources management as unnecessary.

Variety of work

Another point worthy of comment arises from the idea espoused by a number of writers that there is an inherent variety of work in construction that avoids the monotony frequently associated with other manufacturing industries. The general argument is that this variety, along with a visually-evident sense of achievement upon completion of the work, leads to a generally more content workforce and that this can reduce the requirement for specialist HRM. This suggestion must, at least, be questionable. Bricklayers lay bricks and asphalt roofers lay asphalt roofs regardless of which site they happen to be working on. The conditions on site may vary as (most likely) will the people that individuals meet at their place of work, but the work remains essentially the same. If, at some time in the future, we see the introduction of multi-skilled operatives, such as in the USA, then variety in terms of work will become a reality. However, this in itself will not reduce the need for human resources specialists as the motivation of workers is not their concern in any direct sense.

The attitudes and education of construction managers

Traditionally, managers in the construction industry have 'come up through the trades', that is to say they have begun their working lives as apprentice bricklayers, carpenters, joiners, etc. As time went by they would rise to the position of foreman, then general foreman, followed by promotion to site manager and perhaps later to a position of area manager or some other rank within the higher echelons of the firm. One result of this technical education combined with years of experience on the job has been that construction managers tend to be fiercely practical men. Perhaps more than in any other major industry, the management have a propensity to concentrate on the performance of work and the betterment of problems associated with that aspect of business.

Although recent years have begun to show a change, particularly with the increase in full time academic courses aimed at educating strategic/corporate managers for the industry, there is still a substantial element of the 'old school' in positions of responsibility. The attitudes adopted by what we might call the traditionalists are frequently 'exhibited in a reluctance to accept other expertise unless it is unavoidable and demonstrably cost effective' (Rowlands, 1991). Whilst these managers may accept that an accountant or lawyer meet these conditions, it is less likely that they will see HRM in the same light.

Coupled with the points raised above is the desire on the part of those promoted from site to office to retain personal links with former associates from site. This is seen as being part of 'the management' that cares for the workforce by being accessible to their requests and/or complaints and the source of their benefits. Whilst this human face of management is commendable in spirit, it is not always compatible with the taking of objective decisions regarding the running of the company and may actively impinge on, or inhibit, formal HRM.

Review

Human beings are a basic resource of organizations operating in the construction industry. Unlike machines their performance is not predictable and they do not exist in a vacuum. In other words, humans are affected by and have an effect on their environment. The survival of firms depends to a great extent on their ability to interact with the environment.

Beyond survival, success is dependent upon the ability of management to achieve optimum fit between the culture and objectives of the organization and the abilities and expectations of the individual (all of which are influenced by the environment).

One way in which firms can gain an understanding of the potential impact of the environment is through environmental scanning. This involves the identification and analysis of current environmental trends to evaluate their influence on human resources management in the future.

Apart from understanding the relationship between HRM and the environment, we also need to consider the relationship between HRM and the organization. It is vital that human resources managers understand the goals of the organization so that they can begin to integrate both corporate and individual desires. In a construction context, the size of the firm will affect the nature and significance of HRM.

In particular, the following factors are likely to affect the role of human resources managers.

- Changing technology leading to job redesign, recruitment and training; this will require an appreciation of the general culture of the firm.

35

- The wide range of problems and situations likely to arise from the interaction of job performance requirements and the individual characteristics of employees.

HRM is required at each stage in the life-cycle of an organization.

HRM has traditionally been undervalued and therefore underdeveloped in the construction industry. A number of constraints can be identified as responsible for this situation:

- the high cost of HRM when most firms in the industry are small
- the fragmented nature of the industry
- mobility of the workforce
- the shallow management structure generally found at the building site level
- subcontracting and the use of casual labour
- the attitudes and education of construction managers from a trades background

Despite the industry's apparent reluctance to employ specialist managers, changes are inevitable and can already be witnessed in the larger successful construction companies. The traditional 'personal' approach to management is considered by today's industry leaders as limiting. The methods are not considered to be dynamic and cannot evolve to meet the developing challenges in the industry. Furthermore there is a good deal of doubt as to its use in terms of retaining goodwill.

Changes in education and alternative job opportunities are forcing a challenge on traditional methods and assumptions. It may well be that firms with no formal approach to HRM, i.e. where reliance is placed on expediency and *ad hoc* approaches to (variable) decision making, will encounter an increasing sense of grievance in the workforce.

In the future HRM will assume a greater importance in the construction industry simply because so much construction work is labour-intensive and manpower costs are high in relation to total costs. An increasing awareness that people are the most important asset that an organization has and that their effective management must be a key component of a manager's job is likely to lead to an increase in the employment of human resources managers and tools in the future.

Questions

1. Discuss the benefits that can accrue from undertaking environmental scanning and explain why construction firms are less likely than firms in other industries to employ this tool.
2. Discuss how organizational factors can affect the internal environment of a firm with consequences for HRM.
3. Explain why HRM has traditionally been undervalued and underdeveloped in the construction industry.

Bibliography

Armstrong, M. (1988) *A Handbook of Personnel Management Practice* (3rd Edition), Kogan Page, London

Brech, E. F. L. (1975) *The Principles and Practice of Management*, Longman, London

Central Statistical Office (1991) *United Kingdom National Accounts* (1991 Edition), HMSO

Chandler, A. D. (1962) *Strategy and Structure*, M.I.T. Press

Department of the Environment (1992) *Housing and Construction Statistics 1981–1991 Gt Britain*, HMSO

Foulkes, F. K. (1986) *Strategic Human Resources Management*, Prentice-Hall, Englewood Cliffs

Flamholtz, E. G., Randall, Y. and Sackman, S. (1986) *Future Directions of Human Resource Management*, Institute of Industrial Relations, University of California, Los Angeles

Fryer, Barry (1990) *The Practice of Construction Management* (2nd Edition), Blackwell Scientific Publications, Oxford

Hall, D. T. and Goodale, J. G. (1986) *Human Resource Management – Strategy, Design and Implementation*, Scott, Foresman & Co., Glenview.

Handy, C. B. (1985) *Understanding Organisations*, Penguin, London

Heisler, W. J., Jones, W. D. and Benham, P. O. Jr. (1988) *Managing Human Resources Issues – Confronting Challenges and Choosing Options*, Jossey-Bass, San Francisco.

Heneman, H. G., Scwab, D. B., Fossum, J. A. and Dyer, L. D. (1986) *Personnel/Human Resource Management* (3rd Edition), Irwin, Homewood, Illinois

Hillebrandt, Patricia M. (1985) *Economic Theory and the Construction Industry* (2nd Edition), Macmillan, London

Kossen, Stan (1983) *The Human Size of Organizations* (3rd Edition), Harper & Row, New York

Langford, D. A. and Hancock, M. R. (1991) *Management Practice Module MP1 – Human Resources for Construction*, University of Bath

NEDO (1976) *Construction in the Early 1980s*, HMSO

NEDO (1978) *How Flexible is Construction?* NEDO, London

Newcombe R. (1991) *Management Principles Module MT1 – Management – The Macro Level*, University of Bath

Newcombe, R., Langford, D. and Fellows, R. (1990) *Construction Management 1 – Organisation Systems*, Mitchell, London

Pieper, R. (ed) (1990) *Human Resource Management: an International Comparison*, Walter de Gruyter, Berlin

Porter, M. E. (1985) *Competitive Advantage – Creating and Sustaining Superior Performance*, The Free Press, New York

Rowland, V. (1991) *Human Resources for Construction*, unpublished MSc paper, University of Bath

Sikula, A. F. (1975) *Personnel Administration and Human Resources Management*, Wiley, London

Thomason, G. (1981) *A Textbook of Personnel Management* (4th Edition), Institute of Personnel Management, London

Trist, E. L. and Bamforth, K. W. Some social and psychological consequences of the Longwall methods of coal getting, *Human Relations*, **4** (1951), 3 -38; reprinted in D. S. Pugh (1971) (ed), *Organization Theory*, Penguin

Woodward, J. (1980) *Industrial Organization: Theory & Practice*, (2nd Edition), Oxford University Press

Chapter 3

External issues affecting human resources management in construction

We saw in Chapter 2 that human resources managers provide a key link between the organization and its environment. The construction industry exists in a turbulent and complex environment which makes planning for the future extremely difficult. Nevertheless, it is precisely such forward planning upon which the effective management of human resources depends.

Demographic, technological and social value changes, along with changes in the nature of organizations, have important effects on HRM and each of these constitutes a part of the organization's external environment (within the broad headings of competitive, economic, political and technological environmental influences).

Human resources managers need to undertake appraisals of the effects of these external influences to ensure that the firm reaps the benefits of present and potential opportunities and is also able to mitigate the problems that arise through factors that are generally beyond their control.

Demographic changes

Strategic managers in business and industrial communities have begun, in the last 15 years, to become more aware of the need to appreciate the influence of demographic changes on their activities.

Working population

In fundamental terms, these are changes that affect the size of the labour force. The most basic measure is the size of the overall population. Consider the United Kingdom from 1951 onwards set out in Table 3.1.

These figures show that the population of the UK has risen every decade since World War II. Initially the rise was at over 5 per cent per annum, but during the 1970s this slowed to less than 1 per cent. Currently the population is increasing at the UK average for the century of 2.5 per cent. Furthermore we can see that the proportion of women in the population is slightly higher than that of men, but that the difference is reducing. Changes take place as a result of a complex interaction of social, political and cultural factors and so attributing reasons for the trends shown above is difficult. For instance, to what extent were two world wars responsible for the lower percentage of males in the population? And how influential was the introduction of oral contraception methods on the declining birthrate of the 1970s? The human resources manager may not be able to gauge the precise causes of change, but must be aware of general trends that have the potential to affect the labour force.

Table 3.1 UK population 1951–2001

Year	Population (in 000s)	Male (%)	Female (%)
1951	50,225	48	52
1961	52,807	48.3	51.7
1971	55,928	48.6	51.4
1981	56,352	48.6	51.4
1991 (est)	57,561	48.8	51.2
2001 (est)	59,174	49.1	50.9

(*Source:* Mintel, *Annual Abstracts*)

The general population information above, whilst interesting, gives us only the crudest of indicators of change. The human resources manager is most concerned with trends concerning employment.

From the figures given in Table 3.2 we can see a sharp decline in the primary and secondary industries and a consequent rise in the proportion of people employed in the tertiary sector. We also note an increase in the number of women relative to men in employment. This last point is of significance to the human resources manager.

Table 3.2 Employment trends in the UK 1950–2000

Year	Total workforce (in 000s)	Employment (%) in:			Employees (%)	
		Agriculture	Manufacturing	Services	Male	Female
1950	23,068	9.3	46.5	44.2	69	31
1960	25,010	7.6	45.4	47.0	66	34
1970	25,675	3.5	46.6	49.9	64	36
1980	26,759	2.8	38.2	59.0	60	40
1990	28,510	2.1	29.1	68.8	56	44
2000 (est)	32,550	1.8	25.4	72.8	54	46

(*Source:* Mintel, *Annual Abstracts*)

Women in the workforce

The subject of women in construction is considered in detail in Chapter 9. Here we are only concerned with the general role of women in the changing face of the employment market and as a significant demographic factor therein.

The figures above have shown that the proportion of women to men in the overall population has steadily reduced in the last 40 years (from 52 per cent to 51.2 per cent) and yet the proportion of women in the workforce has increased (from 31 per cent to 44 per cent) and is forecast to rise by a further 2 per cent before the turn of the century.

The reasons for the increase in (particularly married) women in the labour force include the declining birthrate which began in 1965 and which accelerated during the 1970s. This left many women available for work who had previously been unavailable. Secondly, social and legislative changes have occurred which may also have encouraged married women to work. Legislation such as *The Equal Pay Act (1970)* and the *Sex Discrimination Act (1975)* were aimed at ensuring both better levels of pay for women and greater chances of employment. Social changes have included a general desire for higher living standards and for increased material possessions. Coupled with these has been greater ambition on the part of women resulting from educational changes and a generally higher level of aspiration (in both sexes), particularly during the 1980s as a result of governmental policy. Changes in the employment trends (i.e. away from manual labour towards the service industries), family planning and changing work patterns have all had a part to play in the increasing female workforce.

Age composition of the workforce

Another important demographic factor resulting directly from changes in the birthrate is the age composition of the workforce. The figures given in Table 3.3 have been abstracted from a more detailed breakdown of age groups, but show an interesting trend. Nineteen years has been selected as an average age for entering the workforce (it may actually be lower) to allow for the differing ages at which people leave full-time education.

Table 3.3 Population trend in the UK by age group 1951–2001

Year	Population (%) by age group		
	0–19 years	20–64 years	65+ years
1951	28.9	60.4	10.7
1961	30.1	58.2	11.7
1971	30.9	55.9	13.2
1981	29.0	56.0	15.0
1991 (est)	25.6	41.3	15.7
2001 (est)	26.3	58.1	15.6

(*Source:* Mintel, *Annual Abstracts*)

The first significant factor is the ageing nature of the population. The percentage of population over the age of retirement has increased by almost 50 per cent in the last 40 years. Secondly, we can see that the proportion of the population available for work has dropped dramatically over the same period from 60 per cent to 41 per cent. Thirdly, the post-war baby boom increased the youngsters in the labour market of the 1970s, but since mid-1981 the number has been in decline as a result of the falling birth rates of the 1960s and 1970s.

Effects of demographic change on construction – the problem of definition

How the figures and changes we have considered so far affect the construction industry is complex, as it is difficult to accurately define which of the three major industrial sectors construction falls into. Fellows (1990) considers the idea of construction as being a single industry questionable. Official statistics class construction as being between a tertiary (services and intangible goods) industry and a secondary industry (concerned with the production of tangible goods). Fellows suggests that this perhaps implies a degree of scepticism as to whether construction is really a service or manufacturing industry. The picture is even more complicated if we consider that there are people working in associated areas in the primary sector (i.e. extraction industries such as quarrying).

Standard forms of business information

It is, of course, impossible to take measurements and to detect changes in construction if we do not have clearly defined boundaries for statistics relating to construction. For our purposes we will adopt the following Standard Industrial Classification (SIC) prepared in the UK by the Central Statistical Office (CSO) (1968 and 1980):

1968 Standard Industrial Classification 500: Construction. Erecting and repairing buildings of all types. Constructing and repairing roads and bridges; erecting steel and reinforced concrete structures; other civil engineering work such as laying sewers, gas or water mains, and electricity cables, erecting overhead lines and line supports and aerial masts, extracting coal from opencast workings, etc. The building and civil engineering establishments of government departments, local authorities and New Town Corporations and Commissions are included. On-site industrialized building is also included. Establishments specializing in demolition work or in sections of construction work such as asphalting, electrical wiring, flooring, glazing, installing heating and ventilating apparatus, painting, plastering, plumbing, roofing. The hiring of contractors' plant and scaffolding is included.

1980 Standard Industrial Classification Division 5 Class 50: Construction

Group 500/5000 General construction and demolition work. Establishments engaged in building and civil engineering work, not sufficiently specialized to be classified elsewhere in Division 5, and demolition work. Direct labour establishments of local authorities and government departments are included.

Group 501/5010 Construction and repair of buildings. Establishments engaged in the construction, improvements and repair of both residential and non-residential buildings, including specialists engaged in sections of construction and repair work such as bricklaying, building maintenance and restoration, carpentry, roofing, scaffolding and the erection of steel and concrete structures for buildings.

Group 502/5020 Civil engineering. Construction of roads, car parks, railways, airport runways, bridges and tunnels. Hydraulic engineering, e.g. dams, reservoirs, harbours, rivers and canals. Irrigation and land drainage systems. Laying of pipe-lines, sewers, gas and water mains and electricity cables. Construction of overhead lines, line supports and aerial towers. Construction of fixed concrete oil production platforms. Construction work at oil refineries, steelworks, electricity and gas installations and other large sites. Shaft drilling and mine sinking. Layout of parks and sports grounds. (Contractors responsible for the design, construction and commissioning of complete plants are classified to heading 3246. Manufacture of construction steelwork is classified to heading 3204. The treatment of installation work is described in the introduction to the SIC.)

Group 503/5030 Installation of fixtures and fittings. Establishments engaged in the installation of fixtures and fittings, including gas fittings, plumbing, heating and ventilation plant, sound and heat insulation. Electrical fixtures and fittings.

Group 504/5040 Building completion work. Establishments specializing in building completion work such as painting and decoration, glazing, plastering, tiling, on-site joinery and carpentry, flooring (including parquet floor laying), installation of fireplaces, etc. (Builders' joinery and carpentry manufacture is classified under heading 4630; shop and office fitting under heading 4672.)

The definition has changed somewhat between the 1968 and 1980 SIC definitions. Notably, construction no longer includes:

* open cast coal mining
* building and civil engineering establishments of government departments, local authorities and New Town Corporation and Commissions; however, direct labour organisations of local authorities and government departments remain included
* design and building contractors' activities and manufacture of construction steelwork which are part of engineering (Division 3 of SIC)

Table 3.4 Present and former industrial classifications

	Present divisions	Former orders
0	Agriculture, forestry and fishing	I
1	Energy and water supply industries	
2	Extraction of minerals and ores other than fuels; manufacture of metals, mineral products and chemicals	II (MLH 101 and 104), IV, XXI
3	Metal goods, engineering and vehicles industries	II (MLH 102, 103, 109), V, VI, XVI
4	Other manufacturing industries	VII to XII inclusive
5	Construction	III, XIII to XV, XVII–XIX
6	Distribution, hotels and catering, repairs	XX
7	Transport and communication	XXIII, XXVI (MLH 884–888, 894, 895)
8	Banking, finance, insurance business services and leasing	XXII
9	Other services	XXIV, XXV (MLH 871, 873)
		XXV (remainder), XXVI (remainder), XXVII

(*Source:* Fellows, R. F., *Construction in the Economy: Module CE2 – Construction Project Economics*)

- joinery and carpentry manufacture together with shop and office fitting which are classified as other manufacturing industries (Division 4)

If we look at the present division and broad classification of industries as set out in the SIC document, we can begin to identify construction amongst a broad classification of industries as in Table 3.4.

Related classifications

1. There are several other classifications both national and international, which may be used in conjunction with the industrial classification. These include, for example, the classification of occupations, which relates to the jobs performed by individual workers rather than to the industry in which they work. The workers classified to a particular industry will fall into a number of different categories of an occupational classification and similarly the workers in some occupations may be found in many different industries.

2. A second classification, used mainly in the national accounts, is that by institutional sector. This groups units according to their organization or ownership, i.e. it distinguishes between unincorporated businesses, companies, public corporations (nationalized industries), central or local government. The industrial classification does not make such distinctions but brings together units engaged in similar activities, irrespective of ownership.

3. A third classification is that of products – often a list of individual products or groups of similar products according to the industries in which they are principally produced. The number of entries depends on how far it is desired to go in the separate identification of products usually produced in the same industry and for similar purposes but varying in the materials used, quality, size, shape, etc. The amount of detail needed for statistical purposes is much less than would be required for, say, a manufacturer's catalogue. The classification can also be extended to cover not only the production of goods but distribution, transport and other service industries though the number of different types of service included will normally be much less than the number of different products of production industries.

4. The alphabetical list of industries and their typical products included in Indexes to the Standard Industrial Classification (Revised 1980) is an example. Strictly, statistics produced on this basis do not measure the full production of an item. Each product or service in general is classified only to one industry – that in which it is mainly produced. In practice many units produce not only the goods or services which are principal products of the industries to which the units are classified but also products mainly produced in other industries. As a result, the total production of the units classified to a particular industry will not necessarily represent the total production of all the items listed as principal products of that industry.

5. At the international level, the United Nations has published lists of principal products under each ISIC industry and an International Classification of Goods

and Services (ICGS) in which the products of each industry group are further classified into subgroups. NACE is supported by the Nomenclature Commune des Produits Industrials (NIPRO) which provides detailed lists of products under each industry but covers only that section of NACE dealing with manufacturing industry.

6. The product lists associated with industrial activity classifications contrast with the classifications for recording imports and exports. The United Nations Standard International Trade Classification (SITC) and the Customs Co-operation Council's Nomenclature (CCCN) with which it is correlated have been widely adopted throughout the world as the basis for national classifications both for tariff and trade statistics purposes. They are used in this way in the United Kingdom and in the rest of the EEC. But in each case the SITC headings have been further subdivided to provide more detail, mainly related to differences in quality, size, etc.

7. The classification is applied to units on the basis of their principal activity. Frequently no problem arises in defining the relevant unit. This will be the case, for example, where the unit is situated at a single address, is mainly engaged in a single activity (or group of activities falling within the same heading of the classification) and is operated as a separate entity for the purposes of record-keeping and accounting (including VAT returns).

8. In many cases, however, the situation is more complex. Many units produce, sell or provide a variety of goods or services, some of which are considered to be appropriate to one heading of the classification and some to another. In such cases the unit as a whole is classified according to the type of goods or services which make up the greater part of its production, sales or service. In governmental inquiries a unit's classification is made on the basis of information relating to the unit which is supplied on the particular return or which is already available from other inquiries. In some cases classification may be based on the unit's own description of its principal activity.

9. The units to be classified may vary according to the type of data being collected and the purposes for which the data are required. In view of this it is important that when analyses by industry are prepared the statistical unit used for the inquiry concerned should be defined. Sometimes, in order to secure compatibility between the results of different inquiries, it may be necessary to give some units more than one classification. This may occur, for example, when a unit at one address can provide a limited range of data, say, employment but for the provision of more comprehensive data has to be combined with another unit, or units, under the same ownership which are mainly engaged in a different activity.

10. The units used in some of the main regular statistical inquiries are as follows:

(a) *Production inquiries.* Information is collected on a wide range of subjects, e.g. production or sales of goods, number of employees, wages and salaries paid, purchases of goods and services from other industries, stocks and fixed capital formation. As far as possible the unit in respect of which such information is collected is one whose activities fall within a single activity heading of the classification and which is situated at a single geographical location. Frequently, however, the information required cannot be provided on this basis but relates to two or more locations and a mixture of activities. The statistical unit, usually termed 'the establishment', is therefore defined as the smallest unit for which the information normally required in these inquiries can be provided.

(b) *Inquiries into the distributive and service trades.* Many of the units in these trades consist of independent shops or other premises for which no problems arise. But an increasing proportion of total turnover, particularly in retail distribution, is accounted for by companies and other organisations with numerous branches and with centralized purchasing and stockholding arrangements. In these cases the unit for which much of the data required can be obtained is necessarily the organization as a whole, although some data, such as sales or numbers employed, may be available for the separate branches. The VAT unit is playing an increasing role in the collection of data in respect of these industries.

(c) *Census of employment.* Within each organization each address at which there are employees is generally regarded as the unit for classification purposes. Normally the whole unit is classified according to the major activity carried out at the address; if however there are two or more quite distinct activities for which separate information can be provided then each is classified separately. The units are, if necessary, given more than one classification, that is according to the major activity of the unit itself and also the major activities of larger entities of which the census unit is a part, and which constitutes the statistical unit for other inquiries.

11. Many businesses produce for their own use goods or services which are normally produced for sale by other industries. Typical examples are the manufacture or printing of containers for their own products, the generation of electricity, research, the repair and maintenance of buildings and equipment, and occasionally the construction or extension of buildings. If the relevant range of information is available these ancillary activities are treated as separate units and classified according to their own activity. But frequently the necessary data is not available, and the ancillary activities cannot be distinguished for most statistical purposes. Where ancillary activities are carried out at separate locations it may be possible to treat them as units for a limited range of data, e.g. for the census of employment.

12. The sales staff of a manufacturing business are not treated as a separate establishment unless separate wholesale or retail outlets are operated and complete accounts are kept for them.

13. The classification of head offices or other central offices at separate locations from those at which the main activities of the business are carried out is determined by the nature of the activities of the units which they serve. If those units are all classified to the same industry there is no problem; but if they are classified to two or more different industries, the classification of the head office, etc. is that of the largest part of the activities. Exceptionally, if the activities are very diffuse, the central office is classified in heading 8396

– central offices not allocable elsewhere – which is also used for the UK offices of businesses operating mainly abroad.

14. Although the classification is designed primarily for use in connection with units such as those discussed in paragraph 10 above it can also be used for other units. For example, it is used by the Inland Revenue for industrial analysis of income tax (schedule D) and corporation tax assessments. In these cases the unit is that which is liable to the tax in question, in many cases a number of establishments engaged in different activities. The data aggregated under a particular activity heading will inevitably relate to a more heterogeneous group of activities than data based on establishments or similar units. Another example is the use of the classification in statistics of financial flows. Here the unit is normally the company or group of companies under the same ownership. These units are even more heterogeneous than those used for tax statistics. As the units increase in size and diversity of activities, classification to a single industry heading at the most detailed level becomes misleading. An abridged version of the classification may therefore be employed in such cases, using broader groupings based on the Groups, Classes or even combinations of Classes of the classification.

15. At the other end of the scale the classification is used in data relating to unemployment, sickness, industrial injuries, etc. where the units are individuals who are classified according to the industry in which they are, or were last, employed.

16. The classification may be used from time to time as a convenient way of defining the scope of legislation relating to particular taxes, subsidies, capital grants, health and safety regulations, etc.

Related statistics

As we have seen there have been major shifts in employment between the primary, secondary and tertiary sectors in the post-war period. In recent years there has also tended to be short-run shifts in the distribution of employment. According to statistics compiled by the International Labour Office, the number of people employed in the UK rose by 11.6 per cent during the period 1982–1989. In that same period, employment in primary industries, such as agriculture and mining/quarrying, fell sharply and stood at 89.5 per cent and 56 per cent respectively of their levels in 1982. The manufacturing (secondary) sector also fell, ending the decade at 91 per cent of its earlier, size but the services sectors all showed considerable growth. Trade restaurants and hotels were up by 15 per cent (projected to rise above their 1982 level by an astonishing 387 per cent by 1991), Financing, Insurance, etc. grew by 50 per cent and Personal Services by nearly 23 per cent. For the period 1982–1991 Construction shows an increase of only 4 per cent suggesting a fairly stable state of affairs but this overall figure masks a series of fluctuations during the period, as Table 3.5 shows.

The construction boom of the late 1980s is clearly evident here as is the beginning of the current economic recession. Unfortunately, more recent figures are not available at the time of writing.

What the figures in Table 3.5 do show is the sensitivity of construction to the general economy. They also tend to indicate that the construction industry falls somewhere between the secondary and tertiary industries which have (unlike construction) both shown stable employment trends (in one direction or the other).

Table 3.5 UK employment within construction 1982–1991

Year	Total workforce (in 000s)	Year-on-year change (%)	
1982	1,465		
1983	1,453	−1	
1984	1,501	+3	
1985	1,492	−1	*Overall change*
1986	1,476	−1	*1982–1991*
1987	1,551	+5	
1988	1,640	+6	
1989	1,805	+10	+4%
1990	1,703	−6	
1991	1,526	−12	

(*Source:* International Labour Office, Geneva, *Yearbook of Labour Statistics 1992* and Department of the Environment, *The Housing & Construction Statistics 1981–1991*)

Technological change

Technology is the key to the understanding of what takes place in the 'black box' that converts inputs to outputs. To a certain extent the level and sophistication of the technology employed by an organization will depend upon two factors that have to be balanced if that firm is to be successful. These factors are managerial philosophy and employee expectations and needs.

Ever since the early days of the industrial revolution there have been arguments over the benefits and disbenefits that arise from the introduction of new technology in the workplace. Those in favour of technological developments argue that it improves standards of living and income levels and increases the amount of leisure time available to people. Those against its introduction contend that machinery displaces human labour and therefore creates unemployment and, furthermore, that the reduced working hours made possible by the technology are not accrued by the workers but by their employers who reduce wages, therefore requiring longer working hours (and therefore higher production levels) from the employees manning the machines.

Whether one believes technological change to be a good thing or a bad thing, there is no doubting its part in promoting economic growth through the raising of productivity.

Recent changes and their effects on construction

A study by the Manpower Services Commission (MSC) in 1977 showed that construction was expected to have a slow rate of technical change in the following

47

ten years and that the significance of any such changes would be little. The reasoning behind this forecast may have been the predicament in which this generally undercapitalised industry finds itself. New technology usually requires large capital investment which only the relatively small number of largest firms can afford. On the other hand, those who don't invest reduce their competitiveness. The MSC did suggest that many of the respondents to their survey may have underestimated the potential of some aspects of technological change; particularly in the area of computer technology. It is worth remembering that micro-computers only appeared on the market for the first time in 1977 and the growth in their sophistication and use was difficult to foresee at the time.

The construction industry has not remained unaffected by the microelectronics revolution. In the 1990s computer-aided design, computerized production of bills of quantities, expert systems and project scheduling and monitoring packages have become commonplace contributions to the modern approach to work in the industry. Alongside these specifically-construction related areas are the general changes brought about by the introduction of computers in terms of clerical and administrative work, e.g. payroll and other records, secretarial work, etc.

The effects of technical change, whilst rather less, have also been felt on the building site. Construction has traditionally been characterized by craft-based work requiring highly skilled operatives with long periods of training/apprenticeship. The craftsmen, be they bricklayer, carpenter, joiner, metalworker or plasterer, were wholly responsible for the quality of the finished product and their workshop(s) and the building site provided an informal environment for social interaction and communication about the job in hand. The increasing size and nature of industrialized processes, brought on by technical change, has led to an increased division of labour in terms of work and its geographical location. Many traditional site-based activities now take place in factories, using automated machinery manned by lesser skilled operatives. There are important implications for motivation (see Chapter 4) as a result of these changes.

We must, however, be careful not to assume that a particular technological change will result in a certain attitude towards work. As we will see in Chapter 4, human resources managers can help employees to overcome difficulties in coping with change through organizational development programmes.

The impact of technology on skills

In his seminal work, Braverman (1974) argued that the introduction of modern technology under the capitalist system leads to a general deskilling of the labour force whether they be craftsmen or clerical workers. There is not time here to investigate the extensive literature that has arisen both in support of and objection to this idea. We might however suggest a somewhat less radical line without fear of contradiction. New technologies introduced to the industry bring about changes in skill requirements and the profile of the workforce. New jobs are created, whilst some existing jobs become redundant. Furthermore the impact of the changes affect the semi-skilled, skilled and clerical worker alike. According to Heisler et al. (1988), professional workers are increasingly at risk. They also state that administrators and managers 'incur the indirect effects of changing technologies. As rank-and-file employees are eliminated or reclassified and repositioned elsewhere, administrative and managerial positions are reduced as well to create more efficient and less costly operations.'

The implications of change for human resources managers

If the problem of human obsolescence is to be avoided, an important implication arises for the Human Resources Manager in the face of changing technology. Those affected will need to be retrained. Skilled lathe operators fully cognisant of the setting-up of a human controlled machine may find themselves at a loss when faced with a computer-operated, numerically-controlled machine. Quantity surveyors, who throughout their careers have produced bills of quantities by the traditional measurement and 'working up' method, are likely to be unused to computerized bill production. Perhaps of greater significance to this example is that the technicians traditionally employed to 'work up' dimensions have become completely unnecessary. Redeployment requires retraining if these employees are not simply to be lost.

As we have witnessed with the prolific development of microcomputing, technological change occurs rather faster than individual human skills and knowledge. Because of this, human resources managers need to be pro-active in formulating plans for managing such change. Considerations in developing such strategies will include the economic implications of investment in retraining and education as opposed to enforcing redundancy or reducing the workforce through natural wastage.

Technological change will also have an effect on management styles. Decision-making and communication are central here. As the use of information technology becomes increasingly commonplace, organizations should tend to become less bureaucratic in their structure with a smaller number of levels in the command chain. According to this line of argument, work groups will become smaller and, as a result, both the need and desire for increased participation (if not autonomy) in decision-making will arise. In this situation the authoritarian style of management and the traditional hierarchical organization structure become less desirable. On the other hand it might be said that in the construction industry, changing technology leads to new difficulties in terms of the timing and choice of communication method and authority. In an industry where complex ideas have to be communicated between designers and other consultants, designers and site personnel, site personnel and their physically removed area or head office, new technology can lead to a growing interdependency between personnel. As the level and rate of interaction between the participants increases there is a greater possibility for communication confusion and conflict arising therefrom.

Another area of potential difficulty lies in the shorter periods of time for older and experienced personnel to adapt to fast changing technology. Today's graduates are much more in tune with modern methods than those who graduated just ten years ago. Human resources managers must realize the potential for tension and conflict that exists through the perception of threats to the authority, power and prestige of the older employees.

As mentioned at the beginning of this section, for technological change to be successful in the organization's terms and beneficial to the employees of the organization then human resources managers need to strike a carefully considered balance between the needs of the two. This can be achieved by adopting a planned framework sensitive to the needs of both.

New work patterns

Construction is characterized by the large number of small firms that operate along with the relative handful of large companies. Although many of these firms will have worked as contractors for the public sector the majority of people working in the industry have actually been employed in the private sector. The exceptions to this rule have been those employed either by a central government agency, such as the Property Services Agency (PSA) or by Local Authorities. In the latter case employment is either in a professional capacity acting on behalf of the Authority in dealing with private contractors, or as publicly employed contractors, such as Direct Labour Organizations.

The 1980s saw the dismantling of the PSA and many DLOs alongside a significant reduction in public sector employment (of all kinds). This was a direct result of governmental policy and as such is potentially reversible. Much less scope for such flexibility is found when the employment market is altered as a result of economic change. The traditional employment system simply cannot cope with rapid economic change and needs to meet the requirements of the market. In other words, increasing competition for work requires labour costs to be kept down.

The rise of subcontracting

The private sector construction industry adopted new employment practices in response to this problem some time ago through the employment of subcontractors on a project-by-project basis. This practice reduces the wage and overhead bills of main contractors during lean times enabling them to remain in business and competitive, but also has detrimental effects on the level of training. Even before the concentrated political trend of the 1980s, Fellows *et al.* (1983) noted that:

> The operatives are predominantly young, male and casually employed, with a strong craft tradition. In recent years there has been a distinct increase in the practice of subcontracting in all trades in response to fluctuating demand and employment legislation; this has to a large extent frustrated unionization of labour.

Professionals within the industry have not escaped changes in the nature of their employment either. Here, increased efficiency has been brought about through the replacement of traditional routine and mechanical activities by the introduction of new technology. As mentioned earlier, computer-aided design, computerized bill production and expert systems are increasingly common. These examples of modern technology can, within limits, store information that removes the need for even some of the most skilled of professional activities. Expert systems are being used to check the results of human judgement and decision making. As the possibility of error is reduced so greater degrees of authority are being allowed at lower levels in the organization hierarchy. Furthermore, changing technology is making it

Table 3.6 Number of construction operative trainees 1981–1989

Year	Number of trainees
1981	60,700
1982	55,000
1983	49,800
1984	49,200
1985	49,100
1986	48,000
1987	44,300
1988	48,400
1989	46,200

(*Source:* Department of the Environment, *The Housing & Construction Statistics 1981–1991*)

Note: Coverage differs from standard industrial classification of 'construction'; in particular, public authorities' direct labour is excluded, as are steel erection firms, while firms manufacturing wooden industrial components are included.

perfectly feasible that consultants, working from home, can exist independently of the traditional office-based practice.

Smaller functional units: opportunities and threats

The trend in employment is increasingly towards small organizations which, with smaller hierarchies, are better able to maintain flexibility and therefore have a faster response time to both opportunities and threats. The large construction firm typically has a smaller head office than previously and a large number of operating units that act as semi-autonomous cost centres. This kind of structure, coupled with generally wider spans of control, allows faster communication between the strategic level of the firm and its operating core. The consequences of these changes include a reduction in the number of middle management levels, less opportunity for promotion within the firm and lower security of employment.

As employment trends in construction change, there is a concomitant change in the nature of career patterns. This poses problems of adjustment for employees. For those who have worked as 'company men' for long periods in large organizations with abundant resources the move to a smaller firm with reduced security of tenure and scarce resources can prove traumatic. This shock is only marginally less for those who remain within large firms which find it necessary to restructure their organization in ways that place the responsibility for continued employment on the employees. Human resources managers have a key role to play here in smoothing the path of such transition.

Flexible employment patterns are also being established among the traditionally craft-based workers on building sites. Following the American model we now see an increase in the instance of multi-skilled workers. The craft sectionalism that traditionally existed within the industry led to a general weakening of the influence of trade unions in construction. This (along with increasing amounts of prefabrication) paved the way for the emergence of a construction operative with basic skills in a number of areas whose main function is that of component assembler. This may be more efficient, but it can be argued that the outcome of this trend is a general reduction in specialized skills with a consequential reduction in the overall quality of the built environment.

Another effect of the trend towards smaller organizations is a reduction in manpower strategies. Many construction firms consider themselves to be too small to bother with strategic planning of any kind. It is difficult to know whether employment practices in the industry are driven by business opportunities that are taken without previous consideration of manpower needs, but the changes of the last twenty years or so would lead one to suspect that the construction industry responds to economic circumstances in an empirical and pragmatic way. The joint effects of economic recession, uncertain workload and technological change have all contributed to the changes referred to above and these appear to have occurred as a matter of opportunity rather than strategy.

Future concerns of human resources management

During a period of recession, the move towards greater subcontracting may appear attractive to firms. Indeed there is some evidence (from the period of intense construction activity during the boom period in the second half of the 1980s) that an upturn in the economy will not lead to a return of directly employed labour. The changes appear to be of an enduring nature. This might lead us to believe that the industry is now 'leaner and fitter' than before, but there are areas of concern for human resources managers which need to be addressed if the benefits of these changes are to be maintained.

Financial considerations: the employment of subcontractors and the contracting out of professional services have undoubtedly led to short term cost savings. The danger for construction firms arises when there is a high demand for labour. Recession in the economy tends to lead to a net outflow of experiences and trained personnel. The costs of employing those that remain committed to the industry when labour is in high demand may, in the long term, prove more costly than retention of at least a basic directly employed core. Human resources managers will also need to watch demographic changes for clues as to likely supply of staff in the future.

Quality of workmanship and services: although the requirements in terms of workmanship provided by subcontractors are usually written into their contract of employment it can be remarkably difficult to ensure and very expensive to correct. Control of this aspect is made more difficult when labour-only subcontractors employ people on a casual basis (known as 'The Lump'). In these cases a question of competency arises as the workforce is largely unknown, (even to those paying their wages) and is frequently selected not for their proven competence but for their youth and strength which will allow fast completion of the work. Related to this form of employment are problems concerning site safety and insurance cover.

In the case of contracted-out professional services the problems are not so great. People selling these services are generally subject to the rules and regulations of their respective professional institutions. Nevertheless, the need to check the credentials of prospective hired professionals remains.

In the long-term, the problem of quality is a rather different one that relates to training provision. Subcontracting and casual employment practices tend to lead to a reduction of training. Large contracting firms with no directly employed labour force have no reason to provide training as they will argue that they pay for trained labour when they employ subcontractors. If those subcontractors are employing people on a casual basis, they will argue that responsibility for training lies with the individual. Whatever the logic, an untrained workforce will produce poorer quality work than a trained one and, according to the Construction Industry Training Board, there was a steady decline in the number of operatives registered for training during the 1980s (see Table 3.6). No figures are yet available post 1989 but there is no reason to believe that the situation has improved.

Whilst perhaps unable to directly influence the above, human resources managers must consider both access to and the content of existing training courses for personnel. All of the major professional institutions require their members to undertake continuing professional development, but what of the site labour force?

Industrial relations: Changing employment patterns present human resources managers with a different set of concerns in terms of industrial relations. What will the effects of any changes be on the directly employed workforce? Will trade union membership and the negotiating mechanism be affected by the changes? How will the level of trust between employee and employer change?

Atkinson's flexible workforce model

During the 1980s John Atkinson, a Research Fellow at the Institute of Manpower Studies, theorized many of the above points and devised a general model of the emerging 'flexible firm' (Atkinson, 1984). The model was based on research which suggested that firms are looking for three kinds of flexibility:

1. *Functional flexibility* which allows quick and smooth redeployment of employees between activities and tasks. This is consistent with the rise of the multi-skilled operative.
2. *Numerical flexibility* which enables a balance to be achieved between the number of personnel employed and the number actually needed which has been realized through the introduction of subcontracting practices.
3. *Financial flexibility* which not only ensures that employers can hire labour as cheaply as possible but is also aimed at introducing new systems of pay which are assessment-based rather than predicated on the notion of the 'rate for the job'.

Atkinson accepts that there is little that is new in firms aspiring to such flexibility. Where he sees change is in firms seeking to 'build-in' all three of these notions to their basic approach to manning, thereby restructuring the work experience of most workers. The new model that he proposes '. . . involves the break-up of the orthodox hierarchical structure of the firm in such a way that radically different employment policies can be pursued for different groups of worker' (Atkinson, 1984).

The key aspect of Atkinson's model (Figure 3.1) is the increasing division of types of employment that allows the firm numerical flexibility through a process that ranks the needs of the organization in terms of its long-term objectives. The model also divides the labour market into primary and secondary categories.

At the centre of the model is what Atkinson refers to as a core group of full-time permanent career employees. This group is drawn from the primary labour market and typically includes managers, designers, quantity surveyors, engineers and other professionals whose jobs are specific to construction. Employees in this category have the highest level of job security, but are expected to accept a degree of functional flexibility. In the short term this may mean acceptance of reduced demarcation between disciplines and multi-disciplinary working (not as big a problem in construction as in manufacturing industries). The main characteristic of this group is the degree of difficulty faced by the firm in trying to subcontract the work that they do. This is particularly so in the case of those with the requisite managerial skills. Employees in this category are generally protected from medium-term fluctuations in the market.

In the secondary labour market, employment is subdivided into what are referred to as the first and second peripheral groups. It is these groups that expand and contract in size according to the state of the market, offering protection of employment to those in the central core.

The first peripheral group consists of clerical staff who are effectively offered jobs rather than careers. They are full-time employees, but their work is not

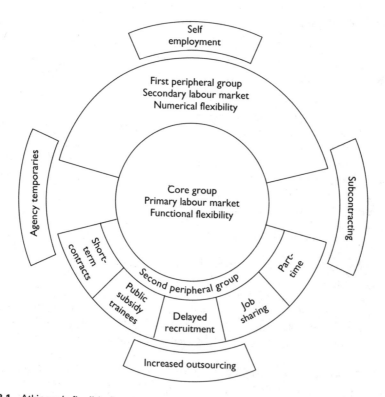

Fig 3.1 Atkinson's flexible firm model. (*After* Atkinson, J., Emerging UK Work Patterns, in *Flexible Manning – The Way Ahead*)

construction- or firm-specific. Jobs in this category tend to be less skilled than those in the core and consist of a narrow range of tasks. Firms will often adopt a recruitment strategy for this group that targets women and encourages a relatively high level of labour turnover. This allows untroubled and speedy adjustment to changes and uncertainty in the construction services market.

The group of employees most exposed to market changes are those in the second peripheral group. Here, employment is often on a part-time or short-term contract basis. Job sharing and the hiring of trainees who attract public subsidies are also found among this group. All of these forms of employment maximize the firm's numerical flexibility (and minimize the firm's commitment to the worker).

Atkinson also identifies a group of workers who are very common in construction: subcontractors. He refers to these as external groups and describes their work as being 'not at all firm-specific – because, for example, they are specialized' (1984). Until the mid-1970s subcontracting did tend to fall into such a categorization but, today, most contractors (in the UK at least) subcontract virtually all of the site-based building operations. In a project-based industry where work is traditionally obtained by competitive means and where future workloads are therefore difficult to forecast, this is not particularly surprising. As Atkinson states, 'This not only permits great numerical flexibility (the firm deciding precisely how much of a particular service it may need at any time), but it also encourages greater functional flexibility than direct employment (as a result of the greater commitment of the self-employed to getting the job done, the greater specialization of subcontractors, or the relative powerlessness of the worker in this context, according to your taste).' (Atkinson, 1984).

Social values

The development of management thinking and writing has tended to reflect a growing awareness of changes in social needs and values. In terms of writing (if not necessarily in practice) we have come a long way since the Classical School of Thought saw the management of people as being similar to the management of machines. Employers have learned, over the years, that humans do not act as neatly programmed automatons: that their behaviour is subject to the influences of sentiment and emotion and that these reactions are founded in the basic social values that are inculcated throughout a person's life through experience and cultural background. To a great extent, good management requires an understanding of the value basis of human behaviour.

Hofstede (1980) defines values as 'a broad tendency to prefer certain states of affairs over others.' He also states that, as values are programmed early in our lives, they are non-rational and that most people hold several conflicting values at the same time. He cites freedom and equality as examples. Managers are immediately faced with a difficulty here, because if employees come from identical backgrounds and experience (a virtual impossibility) they may still hold conflicting views about what is desirable. How then is it possible to mould the values of the individual with those of the organization? Indeed, how is an organization able to hold a single set of values? The answer seems to lie in the acceptability of collectivity. In other words, we accept the preferences of the majority. For the human resources manager the key to success is in understanding and appreciating which preferences hold sway at any given time.

Individualism

An important factor in determining collective social values is the degree of individualism held by the group. Research (Hofstede, 1980) has shown that wealth (GNP per capita), geographical latitude and organization size give an accurate indication of the level of individualism one might expect to find. Generally speaking, wealthier, colder countries with large organizations tend to place more emphasis on the individual. Additional factors that affect this are historical and traditional elements. If we take the UK as an example we can see that the relative wealth (in international terms), cool climate and large firms (generally) indicate a society that places considerable emphasis on the individual. If we then consider the fragmented trade union system that exists in construction, where crafts have traditionally competed with each other and have difficulty in conducting joint action, we can confidently presume that workers in the UK construction industry have an individualistic value system.

Numerous connotations arise from the notion of individualism in the collective value system and these must be understood if managers are to be successful. Hofstede lists many of these, but only a selection of them needs to be given here to demonstrate the point. Furthermore, if we compare these connotations alongside those that identify with societies where individualism is not considered so important we begin to understand the significance of this aspect of social values in management. Figure 3.2 shows a comparison between societies with high and low individualistic values: if we consider the UK to be represented by the High Individualism column, it is not difficult to think of Japan when looking at the Low Individualism list.

Work as a central life interest

As lower order needs (identified by Maslow as physiological and social) are assuaged, so money becomes less of a motivator in the work situation. For the majority in the so-called developed countries these needs have already been met and the motivation for work changes. Increasingly people see work as a central interest in their lives for either sociological or psychological reasons. According to Kahle and Timmer (1983) this is especially so for women, older people and younger professionals. Ferguson (1982) suggests that people are increasingly concerned with lifestyle, psychological quality, social meaningfulness and with corporate accountability. In reviewing these findings, Flamholtz *et al.* (1986) state that:

> These trends imply that work will be decreasingly valued exclusively as a means of financial support and that individuals will increasingly turn to their work as a source of personal identification, social relationships, and community responsibility. A firm's ability to attract qualified workers is becoming increasingly related to the personal enrichment it offers employees and to its image as a responsible executor of communal and environmental resources.

Whilst the suggestions above may be true for the groups specified, construction is an industry dominated by (generally young) males. It is also an industry in which there are sharp differences between managers and/or professionals and site operatives. Although his study related to international differences in work-related values, we can again look at Hofstede's work in assessing the likely implications of work centrality for construction. Hofstede looked at the area of masculinity and work centrality and found that in terms of psychological values 'Regardless of country or occupation, "challenge" and "earnings" tend to be the two work facets whose satisfaction is most strongly correlated with overall satisfaction' (1980). However, when he investigated the area of sociological values (i.e. employees' social frame of reference and feelings of social desirability) he found that both country and occupation are significant in determining the degree to which work is seen as central to people. In countries such as the UK and Germany maximal job satisfaction is considered important and therefore work is seen as being more central than in countries such as France and Holland where life satisfaction is less connected with job satisfaction. Hofstede also considers that the same reasoning can be applied to occupations, with managers and professionals being more concerned with achieving life satisfaction through work than is the case with lower status employees.

HIGH	LOW
Emotional independence from company	Emotional dependence on company
Small company attractive	Large company attractive
Calculative involvement with company	Moral involvement with company
Managers aspire to leadership and variety	Managers aspire to conformity and orderliness
Managers rate having autonomy more important	Managers rate having security in their position more important
Managers endorse 'modern' points of view in stimulating employee initiative and group activity but . . . individual decisions considered better than group decisions	Managers endorse 'traditional' points of view, not supporting employee initiative and group activity but . . . group decisions considered better than individual decisions
Enjoyment in life appeals to students	Duty in life appeals to students
Pleasure, affection and security as life goals	Duty, expertness and prestige as life goals
Individual initiative is socially encouraged	Individual initiative is socially frowned upon: fatalism
Less acquiescence in response to 'importance' questions	More acquiescence in response to 'importance' questions
People thought of in general terms: universalism	People thought of in terms of in-groups and out-groups: particularism
Need to make specific friendships	Social relations are predetermined in terms of in-groups

Fig 3.2 Connotations for human resources managers of individualism in social values. (*Adapted from* Hofstede, G., *Culture's Consequences*)

Age differences

We are all aware of terms such as 'the generation gap' and of older people puzzling over or complaining about 'the youth of today'. These phrases occur because, just as people's physical conditions change with age, so do their values. Human resources managers need to be aware of the differences that can occur and of their causes. These fall under four main headings:

1. *Maturation* – the changing of values as people grow older. If a manager is dealing at a single point in time, with people who are all of a similar age, then this aspect tends to be irrelevant. Its main relevance is when there are people of varying ages to be considered, or when the workforce as a whole is managed over a period of time.
2. *Seniority* – differences in values not as a result of age, but due to position of seniority in the firm and the consequent levels of commitment, market value to other firms and frustration with the job.
3. *Generation* – values instilled in individuals during their youth which are then carried throughout their lifetime. When living conditions change drastically then these fixed values may also change resulting in differing sets of values being held by different generations.
4. *Zeitgeist* – changes that fundamentally alter everyone's values regardless of age.

Age differences also have implications in terms of the wants and needs of those in respective groupings. We might reasonably consider that younger employees will place more importance on developing their abilities, career advancement, pleasure, personal time and earnings. Older employees are more likely to have job security, social relationships, comfort and co-operation as their priorities.

Different needs of men and women

In a traditionally male-dominated industry like construction, differences in the needs of men and women are perhaps more obvious than in other industries where the degree of dominance is lower.

For women, equality in terms of rewards and promotion is an essential if the industry is to demonstrate an attempt to end its traditional discriminatory practices. The gaining and protection of self-esteem is an important female need. Another important need for women is the flexibility to divide their time between work and family. More flexibility on the part of employers is being demanded. Women's values have changed and they are no longer prepared to make the choice between career and family as the women of their mothers' generation did.

The work environment, both physically and socially, tends to be more important to women than to men. Although building sites are not renowned for their physical comfort, improvements can be made. In terms of the social environment women tend to be more concerned with working conditions and social relations with co-workers than do their male counterparts.

As a greater number of women join the workforce, so the imperative for men to be 'the provider' will wane. Men's needs in terms of wages and promotion will therefore be less significant. Many of today's working men place higher value on increased leisure time than increased wages once they are established in their careers.

Differences in the values of men and women tend to evaporate as they grow older and (according to Lannon, 1977) sex role behaviours become blurred as they reach middle age.

Review

Whilst the internal environment of a firm can (to some extent) be controlled, influences from the external environment can produce different and difficult sets of problems for the survival and well-being of the organization.

In order to reduce the level of turbulence in the firm's external environment, human resources managers need to be aware of the constantly changing nature of the environment and of any trends that arise within and from it. Many aspects of environmental influence seem, at first glance, to be small and insignificant if viewed in isolation. The key to success for human resources managers is in being able to identify the numerous and often subtle changes and interchanges that take place and to maintain an awareness of their potential benefits and threats to the organization.

Demographic changes affect the size and nature of the available workforce. We have seen for instance that there are an increasing number of women available for work and that more people are working in the services sector than ever before. We know that the workforce is ageing and that there have been short run shifts in the distribution of employment in recent years.

The technology available also changes and affects people working in the industry. The introduction of computers in the office and new materials (and methods of handling them) on site have important implications for the organization of work and on the motivation of those employed. Management styles, including the decision-making process and communications, are also affected.

Changes in the economic environment have led to changes in working patterns. This is perhaps most obviously noted in the rise of subcontracting. Changing employment patterns and the desire for greater flexibility, both on the part of employers and employees, bring changing career expectations and have significant effects on training.

Social values are also changing and human resources managers must be aware not only of society's preferences at any given time, but of differing sets of values held by different groups in society. Influences here are the emphasis placed on the individual as opposed to the group, centrality of work in the individual's life and the effects of age and sex on values held.

In order to identify environmental trends, human resources managers need to be both forward and outward looking in their approach.

Questions

1. Explain how demographic and technological change affect the work of the human resources manager, making particular reference to the construction industry.

2. (a) Consider the extent to which Atkinson's model of the flexible firm is pertinent to the construction industry;

   (b) provide a response to the suggestion that the flexibility to which Atkinson refers only benefits the industry's employers and not its employees.

3. Social values form the basis of human behaviour. Describe how this is likely to have a direct bearing on human resources management in a fragmented industry such as construction.

Bibliography

Armstrong, M. (1988) *A Handbook of Personnel Management Practice* (3rd Edition), Kogan Page, London

Atkinson, J. (1984), Emerging UK work patterns, in *Flexible Manning – the way ahead*, IMS Report No. 88, Institute of Manpower Studies, Brighton

Atkinson, J. (1985), The changing corporation, in Clutterbuck, David (ed), *New Patterns of Work*, Gower

Braverman, H. (1974) *Labour and Monopoly Capital*, Free Press, New York.

Burack, E. H. and McNichols, T. J. (1980) Human resource planning: technology, policy, change in Miller, E. L., Burack, E. H. and Albrecht, M. (eds), *Management of Human Resources*, Prentice-Hall, Englewood Cliffs

Cross, M. (1984) Flexibility and integration at the workplace, in *Flexible Manning – the way ahead*, Institute of Manpower Studies, IMS Report No. 88, pp. 38–50.

Fellows, R. F. (1990) *Construction in the Economy: Module CE2 – Construction Project Economics*, University of Bath

Ferguson, T. (1982) *Successful Women: A Comparison of Self-Perceived Factors Influencing Successful Women Born 1910–1915*, Dept. of Education, University of California, Los Angeles. Doctoral Dissertation, Microfiche No. 0385591MC

Flamholtz, E. G., Randall, Y. and Sackman, S. (1986) *Future Directions of Human Resource Management*, Institute of Industrial Relations, University of California, Los Angeles

Foulkes, F. K. (ed). (1986) *Strategic Human Resources Management*, Prentice-Hall, Englewood Cliffs

Fryer, B. (1990) *The Practice of Construction Management* (2nd Edition), Blackwell Scientific Publications, Oxford

Hall, D. T. and Goodale, J. G. (1986) *Human Resource Management – Strategy, Design and Implementation*, Scott, Foresman & Co., Glenview.

Heisler, W. J., Jones, W. D. and Benham, P. O. Jr. (1988) *Managing Human Resources Issues – Confronting Challenges and Choosing Options*, Jossey–Bass, San Francisco.

Hofstede, G. (1980) *Culture's Consequences* (Abridged Edition), Sage Publications

Kahle, L. R. and Timmer, S. G. (1983) A theory and method for studying values, in L. R. Kahle (ed), *Social Values and Social Change*, Praeger, New York

Lannon, J. M. (1977) Male versus female values in management, in *Management International Review*, **17**,1, 9–12

Manpower Services Commission/Training Services Agency (1977) *Training for Skills: A Programme for Action*

Megginson, L. C. (1967) *Personnel Management – A Human Resources Approach*, Richard D. Irwin Inc., Homewood, Illinois

Newcombe, R., Langford, D. A. and Fellows, R. F. (1990), *Construction Management: Organization Systems Vol. I*, Mitchell

Odiorne, George S. (1984) *Strategic Management of Human Resources*, Jossey–Bass Publishers

Peterson, R. B. and Lane, T. (1979), *Systematic Management of Human Resources*, Addison-Wesley Publishing Co., Reading, Mass

Prest, A. R. and Coppock, D. J. (1976) *The UK Economy: A Manual of Applied Economics*, Weidenfeld & Nicolson

Ray, G. F., UK Productivity and Employment in 1991, in *Futures*, April 1978, pp. 91–108.

Toffler, A. (1971), *Future Shock*, Pan Books Ltd

Chapter 4

Organizational behaviour

Modern society abounds with organizations. Whether it is as the most junior member of a youth club or the chairman of a multinational corporation we all belong to an organization of some kind during our lives. It might seem easy in that case to define what we mean by the word, but this does not appear to be the case. Wherever we look in the management literature we find differing definitions of the organization.

The organization

Consider the following few examples:

- Organizations are grand strategies individuals create to achieve objectives that require the effort of many (Argyris, 1965).
- By organization, we mean social units devoted to the attainment of specific goals. In this sense, organization stands for 'complex bureaucratic organization' (Etzioni, 1964).
- An organization is a social device for efficiently accomplishing through group means some stated purposes; it is the equivalent of the blueprint for the design of the machine which is to be created for some practical objective (Katz and Kahn, 1966).
- The process of co-ordinating different activities to carry out planned transactions with the environment (Lawrence and Lorsch, 1967).

The distinguishing features between organizations like the youth club and a major corporation are the purpose of the organization and the need to be formally structured and organized. This need is likely to arise from the complexity of task and size of organization. We might, then, usefully separate business or formal organizations from social or (relatively) informal organizations.

Organizational planning

For an organization to be efficient, effective and robust, it must be able to cope with risk and uncertainty. Rarely can this state of health be achieved by chance so plans must be made. Aims and objectives must be agreed upon and goals and targets established.

Although the need for planning exists in all industries it is particularly vital in construction owing to:

- The complex nature of projects which require the co-ordination of large amounts of human and material inputs.
- The generally competitive methods of gaining work which threaten continuity in an already hostile environment with turbulent markets.

For the human resources manager the initial concern will be at the strategic level. They must first be aware of the products, methods and markets within which the organization intends to (or already does) operate. This information will allow an analysis of likely workload and type and then of the personnel requirements of the organization. This knowledge will also enable suggestions to be made concerning organizational design and structure. The purpose here is to facilitate the co-ordination and direction required to achieve the agreed objectives in as controlled a manner as possible.

The human resources manager's contribution to the planning and design of the organization is vital and can be described at two levels; macro and micro. At the macro level concern is with overall structure. This includes specifying channels of communication, methods of control and the allocation of roles and responsibilities. Because of the difficulties faced by construction firms (mentioned above) it may be that a more federal structure is appropriate to the organization than the traditional bureaucratic hierarchy. In this case it is important that special attention is given to these issues in order to overcome feelings of insecurity concerning employment and position in the firm among the workforce. This last point leads directly to the micro level. Whatever the structure finally decided upon, it will be necessary to ensure that the work to be done tallies with the job description and that employees are motivated. This can be done by providing them with responsibility and a chance to achieve through the use of what skills they have. Additionally, training and development programmes can be established to provide opportunities for learning and promotion.

Organizational development

For successful firms, the planning and design of an organization is not a finite activity that is done and then forgotten. Whether the organization is new or established, it will continue to evolve and be subject to environmental forces that require an action or reaction on the part of the firm if it is to survive. The keyword here is change. Changes commonly faced by construction firms include:

- booms and slumps in the market
- technological change
- company mergers and/or takeovers
- the need to gain economies of scale through the combining of activities
- centralization or decentralization of operations according to whether there is a need to increase or reduce bureaucratic control

In order to cope with changes on such a broad front, many firms have adopted Organizational Development (OD) programmes. OD programmes use a behavioural science approach to systematically analyse and understand the organization's environment and the likely changes arising from it and problems affecting it. The overall aim is to improve the way(s) in which the organization copes with change, and in particular with the human aspects of it such as participation, conflict, communication and interaction.

Unlike Management Development programmes which tend to concentrate on individuals being trained to keep up with technological or other advancements, to prepare for promotion, or simply as morale-building exercises, OD seeks to alter the organization's internal environment or culture.

Central to the success of any exercise concerning change is the acceptability (to the employees) of the individual(s) employed to run the programme. Commonly referred to as the 'Change Agent', they can be external consultants who, as outsiders, are able to view the organization objectively and bring fresh ideas, untainted by the existing culture, to the firm. Difficulties might include resentment on the part of employees to interference from outside and a feeling that those who have the best understanding of the organization (the change agents themselves) are not part of the organization itself. These consultants are usually expensive to employ, but only the very largest construction firms would realistically be able to afford the alternative which is a specialist department within the organization.

Fryer (1990) suggests that a combination of both of the above alternatives, along with the assistance of line managers (to ensure commitment and ideas from those on the inside), is often the best option. This may well be so, but the cost implications cannot be overlooked and it seems more likely that an affordable method might simply be a group committed to the notion of change drawn from senior management in conjunction with members of line management and other influential staff.

The important factors in the success of an OD programme will be the social skills of those preparing and implementing the plan and their ability to convince those who will be affected by it of its validity and desirability. If those affected by the programme perceive a measurable improvement in conditions as a result of OD, then it will be accepted and will stand a good chance of becoming a continuing factor of the organization's life.

The process of OD is firstly to undertake a comprehensive analysis of the organization and to explain the necessity of the exercise to employees. It will be important throughout the process to unlock people's existing attitudes and fears about change. These are expressed in a multitude of ways such as concern for the breaking up of existing work-groups, fear of redundancy, loss of status, the need to learn new skills, transfer to different departments or working for a new boss, etc. Techniques to accomplish this include, team-building and intergroup activities, survey and feedback sessions, education and training programmes for employees, career planning and counselling sessions and consulting the workforce so that they feel a sense of commitment to the process as a whole.

Having persuaded employees to accept the idea of change it is necessary to establish the attitudes and behaviour that will constitute the new culture. Only if the new culture is inculcated will the OD programme prove successful. This can be realized through positive reinforcement from senior management.

Finally, comes the need to monitor and measure the success of the programme against the plan. This feedback will help to support successful aspects and detect further problems in need of attention.

How successful OD programmes are is difficult to determine. The process is concerned with whole organizations and so there are difficulties in carrying out controlled studies. However, we might say that it is an attempt to create a flexible and dynamic organizational form. Given that such organizations survive better in turbulent and hostile environments and where human skills are a critical element of success, it might be reasonable to assume that benefits will accrue to construction companies who engage in them.

Organizational typology

Organizations consist of people doing jobs and holding positions. As a result of the division of labour necessary in organizations, two types of employee can be identified: managers and operatives. Within the category of those classified as managers there is a further division into specialist and functional management.

One useful way of identifying types of organization is by looking at their structure. By this we mean the formal pattern of authority relationships between the people (managers and operatives) and/or departments that form the subsystems of the firm.

Line and staff organizations

Typically, construction organizations are faced with something of a dilemma. How can they structure themselves in a way that allows the harmonious working together of two different kinds of management both of which are necessary for the work of the firm, i.e. general supervisory managers and specialist function managers? This is what Taylor referred to as the division between the brainwork and supervision of production.

The most common solution adopted by the construction industry is the Line and Staff Organization. This is constituted of a vertical chain of command where those holding line positions exercise a formal authority over subordinates in lower positions on the organizational chart. This allows them to unilaterally issue or pass instructions and information down the hierarchy; authority is direct and uncomplicated. Staff usually act in a supporting role to those with line authority. Those holding staff positions are specialists whose advice and support to line management enables functions, deemed necessary by the organization, to be carried out. Authority for those in staff positions is usually limited to the area of their particular expertise.

In construction some individuals perform both line and staff functions. For example a quantity surveyor working in a department of quantity surveyors will have a line relationship with the Chief Surveyor from whom they take orders, but is also likely to have a staff relationship with the firm's site managers and supervisory personnel.

The rudimentary design of a line and staff organization is then quite simple as shown in Figure 4.1, but modern organizations are not as uncomplicated as those referred to by the early management writers. Not only has there been a tendency to expand and diversify, but the peculiar fragmented nature of the construction industry raises other questions.

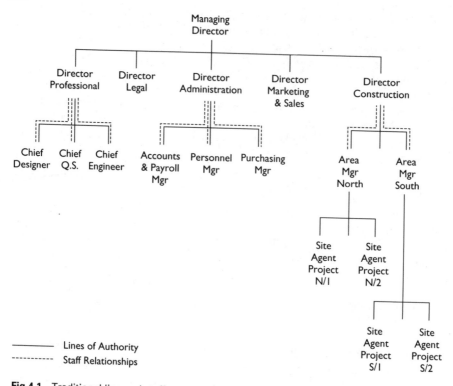

Fig 4.1 Traditional line and staff organization for a construction firm

Firstly, given that people are individually so variable both in character and their competence to supervise others, how many personnel should any individual be expected to supervise and control? Secondly, as firms expand and diversify, how can the organization's structure cope with new and sometimes overlapping areas of expertise? Thirdly, how will the firm deal with possible changes in the geographical location of its operations?

How many personnel should be controlled and co-ordinated by an individual is debatable. A number of writers suggest that previous research has shown a span of control of between six and ten subordinates to be ideal. What few of them mention is the nature of the industry(ies) studied. Construction is comprised of a wide range of organizations with differing structures and operations. Professional consultancies (unless very large) tend towards a flat management structure with the partners directly controlling a relatively large number of staff. Building companies are to some extent divided between their Head Office and site operations. Head Office is likely to be more in line with the traditional concept of a line and staff organization with a restricted number of subordinates to each manager, whilst on site we might expect to observe a flat structure similar to that in the professional consultancy with large numbers of subordinates reporting directly to the site manager.

The traditional line and staff system is based on lines of authority and the functions of the organization, but in the wake of expansion or diversification, construction

firms find it necessary to rethink this structure. Companies expanding t
operations, often in a geographical sense, tend to divide their functions according
to area and to decentralize a number of the administrative functions. This leads
to a degree of autonomy and accountability (especially for profit and costs) within
the local division whilst Head Office retains strategic control. A similar pattern can
be observed when a firm diversifies. Here they are likely to divide the organization
according to product, e.g. housing, commercial, industrial, civil engineering, etc. In
both cases the object is to provide a structure capable of the flexibility and quick
response necessary for effective operation. Both area- and product-based divisions
can be structured in a similar way to the Head Office model (i.e. according to
function) unless the firm both expands and diversifies. In this instance a balanced
compromise will be required between the demands of location bases, product type
and function. Sample organizational structures are depicted in Figure 4.2.

Apart from the difficulties of design noted above, we might also identify other
weak points in this organization type. Central to the line and staff type of
organization are the concepts of function and authority. This of course is only
a model and may well understate or give an oversimplified idea of the actual
relationships between the actors involved. In particular this may be the case in
terms of power, influence and control. A relevant example of this is given by Cascio
& Awad (1981):

> Although a personnel manager is technically a staff member of an organization,
> his or her real authority in deciding on recruitment, transfers, promotions,
> dismissals and the like, in some cases is final.

In this example a staff member (supposedly advisory) has the power, influence
and control of a line manager. Furthermore, as we shall see when considering the
organization structure of a construction project, the usually accepted and understood
definition of line positions may well be questionable as the nature of the authority
held by individuals changes according to organizational type.

Matrix organizations

A distinguishing feature of the construction industry is the individual project
nature of its operations. Here the temporary multi-organization responsible for the
undertaking is likely to be formed especially for, and disbanded immediately after
completion of a building project. Such a situation requires a type of organization
better able to cope with the demands of numerous one-off schemes, particularly
where there is a high level of discretion attached to the roles of individuals within
the team. Such a structure is the Matrix Organization. Underpinning this type of
organization is the idea that the organization of work be kept separate from the
organization of people. In order to ensure that both work and people are managed,
part of the traditional command hierarchy is replaced by a network of vertical and
lateral associations. As shown in Figure 4.3 project managers are responsible for the
management of the work undertaken on individual projects while people managers
are concerned with the development, welfare, qualifications and competence of the
same people in terms of their overall role in the company.

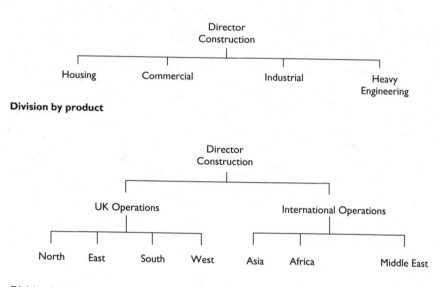

Division by product

Division by geographical area

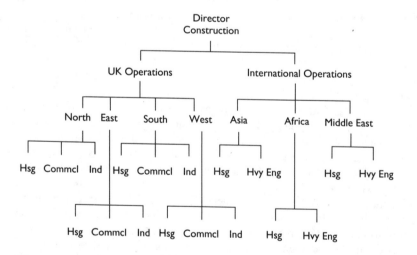

Combined product and geographical area structure

Fig 4.2 Various organization models for the line and staff structured construction firms

The matrix structure was initially adopted both by building firms and professional practices in their efforts to find an alternative to the purely functional approach of the classical school (Taylorism in particular). It was felt that this type of organization would allow both project/task problems and people problems to be dealt with separately, thus avoiding interference by one of the other and with benefits to both aspects of the organization's operations and the employees themselves. Furthermore, a matrix organization ensures that different kinds of task are staffed and accorded an amount of freedom or control commensurate with their type. In construction this might typically range from routine jobs undertaken by technical personnel that require repetitive solutions, through to heuristic tasks that are unique, non-repetitive and with no clearly measurable output such as the work of architects. Employee appraisal is in these cases a three-way affair, involving the project manager, the people manager and the individual person. This helps to alleviate claims of unfair treatment based on clashes of personality or non-understanding of the particular type of work being done.

Difficulties with the matrix organization have arisen through divided loyalties and the feeling by workers that they must try and satisfy two bosses who both have different concerns, strengths and weaknesses. To which manager should the worker show loyalty: the manager of the project or the specialist group or department with whom they identify within the firm? Fortunately some of the problems faced by other industries that have attempted to implement the matrix structure are generally absent in construction. For instance, what Handy refers to as the problem of 'role clarity' (Handy 1985, p. 314), i.e. confusion arising from being a member of a project team for the first time and therefore outside of the typical bureaucratic structure, is not an issue for those in an industry that has always been project-based. Nevertheless, for a matrix to be successful, it is important that the workers be able to discard entrenched practices and departmental allegiances thereby allowing themselves to move freely between their specialist role in the company and their task role on the project.

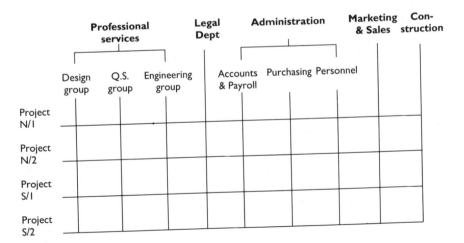

Fig 4.3 Matrix structure for temporary organizations

Project organizations

Whilst both the line and staff and the matrix organizations attempt to cope with the combinations of people within the organizational structure of the firm, we have already seen that construction is essentially project-based. Although the matrix structure was certainly devised as a response to the needs of organizations involved in project-based activities, each project in itself becomes an organization. Generally, in construction, projects can be described as highly complex undertakings requiring regular co-ordination of the work done and co-operation between a wide range of functional specialists, crafts and tradespeople. Upon completion of the project all of these people will return to their own departments (if directly employed) or firms.

The project organization tends to combine the formality of the classical school (line and staff structure) with the lateral and more relaxed communication channels epitomized in the matrix structure. This then allows the project manager to exercise necessary control, whilst at the same time enabling the freedom of discussion required to overcome collective and frequently overlapping problems.

Organizational objectives and goals

Regardless of which sector an organization operates in, the true measure of its success will be in whether or not it satisfies the requirements of those who have an interest (pecuniary or otherwise) in it. The fragmented nature of the construction industry coupled with the belief that 'most industrial organizations have fairly limited objectives' (Fryer, 1990), has increasingly meant that construction organizations have tended to abdicate or ignore any responsibility for the needs of employees. The problem here is that if the managers of firms in the industry are not aware of the aspirations and requirements of their employees they are ill-equipped to gauge whether the firm is performing optimally.

Whether or not organizations are capable of having objectives and goals is a matter of some debate. Fryer (1990) claims, 'The modern view is that an organization is a coalition of people. The organization, being mindless, cannot have goals – only the people in it can.' On the other hand Cascio and Awad (1981) consider 'formal organizations are part of the goal attainment subsystem of the larger society. They are major mechanisms for mobilizing power in the interest of achieving specific objectives.' This they claim supports their assumption that 'organizations have operative goals or role prescriptions within the framework of the higher-level goals of society'. Regardless of the differences in these two stands, they both share a common concern for society. Society being constituted of human beings, it is an easy step to the argument that organizational aims and objectives should insofar as they steer the business of the organization, place at least equal importance on the human resources function as they do on the economic, marketing or production functions.

The goals of an organization are rather more problematic to deal with. These are the beliefs on which decisions are founded and will quite naturally differ according to the values held by the decision maker. In construction, possibly the only goal shared by every member of the organization (and those affected by its activities) is that of survival. Beyond this, almost every level of participant, whether employer, employee, self-employed or client will have different priorities. It is, then, important for the organization's goals to be clear, concise and consistent with an established set of objectives that are, as far as is possible, agreeable to all concerned.

This situation is even more pronounced at the project level where the organization is not usually permanent, but is a Temporary Multi-Organization. Extra difficulties lie here, because the nature of the project organization means that there are many individuals making decisions of one kind or another who each belong to different permanent organizations. As well as the objectives and goals of the project itself, they are confronted with possibly conflicting sets of ideals from their permanent organizations and personally held beliefs.

In establishing its goals the organization must be mindful of the need for a lucid statement of its objectives and of the need for periodic updating; inclusion of new objectives, removal of redundant ones, reappraisal of priorities and, where necessary, changing the overall direction of the organization. How much and how often this updating process is undertaken is critical. As we have already seen, organizations are continually interacting with the environment which is itself continually changing, but care is required. To ignore environmental demands for change will almost certainly result in loss of opportunities, skilled workforce and other inputs. In extreme cases the organization may be forced to cease operating at all. To be over-sensitive to changes in the environment and constantly amending goals can mean possible failure in the achievement of the main objective of the organization and the impression of instability with consequent loss of employees and clients.

Given that the effectiveness of an organization is to a great extent measured by its ability to achieve its stated goals, there is a clear message for managers here. It is essential that the core organizational goals and the likely frequency of them changing, form the yardstick for the human resources policies and plans of the enterprise.

Leadership

The effectiveness of the organization also depends on the ability to integrate the workforce into a well-motivated and productive team that is committed to the completion of projects and overall success of the firm. Undoubtedly, a sense of collective purpose and meaning needs to be achieved among the diversified and fragmented workforce. The problem of fragmentation is further exacerbated by the continually changing composition of the workforce (on a project-by-project basis) and by modern employment practices in the industry such as subcontracting. These difficulties tend to result in the creation of a turbulent 'internal environment' every bit as complex as the organization's external environment. To deal with this requires not just good management, but strong leadership.

Frequently the words management and leadership are considered to be inter-changeable. Whilst they may be synonymous, in fact they are not the same thing. Leaving aside Mintzberg's (1975) definition of management which sees leadership as a managerial role, management consists of dividing up work and then co-ordinating the integration of that work, which is done through others, to achieve the organization's expressed objectives. Management, then, is concerned with the activities of people with prescribed roles working within the organization, i.e. procedures and results.

Leadership has been described in many ways, but all definitions seem to have a common thread. It is that leadership is seen to be the influencing of others to do what one individual wants them to do. Leadership goes beyond the formal activities of managers and introduces a broader interpersonal activity.

Ideally, all managers should be leaders, but there are many leaders who are able to influence others by virtue of their personal charisma and who have no capabilities in terms of other managerial functions. For the purposes of this book we shall borrow from Robbins (1991), and define leadership as the influencing of others to do what an individual with managerial authority wants them to do.

Theories of leadership have evolved from studies of social psychology. Writers have used seemingly endless terms to describe the various styles that exist; open, democratic, people-oriented, participative, closed, autocratic, production-oriented, laissez-faire. Despite the plethora of terms, the best known and developed theories of leadership can be grouped under three broad headings:

1. trait theories
2. behavioural theories
3. contingency theories

The trait approach

Central to this approach is the belief that leaders can be differentiated from their followers through the evaluation and comparison of their physical, mental and psychological characteristics. The idea is that there are certain individual traits that separate not only leaders from followers, but also effective from non-effective leaders. For more than 70 years researchers in the field have investigated human characteristics such as aggressiveness, ambition, decisiveness, dominance, initiative, intelligence and physical features in attempts to prove that leaders are born, not made. This is often referred to as the 'Great Man' theory.

No empirical research has ever managed to substantiate this notion, although a number of features have been identified as consistently present in effective leaders. These include dominance, high levels of energy, intelligence and self-confidence. We must however, be careful not to confuse the correlation between these traits and leaders with the idea that they are definite predictors of leadership.

In fact, other approaches to the study of leadership have highlighted the fallacies and limitations of trait theory. Criticisms are based on the view that trait theory ignores the needs of the followers, fails to clarify the relative importance of various traits and ignores situational factors. Nevertheless, there are still those who subscribe to the theory and in recent years there has been a revival of interest in it.

Behavioural (or style) theories

Between the late 1940s and the mid-1960s management thinkers were increasingly dissatisfied with the trait approach to leadership. Research during this period became more concerned with how leaders should behave rather than with what characteristics they do or should possess.

Researchers hoped that by identifying behaviours that differentiated effective from ineffective leaders they would discover more clear-cut answers to questions about the nature of leadership.

This approach to the subject provided the first major challenge to the trait theorists. Behaviourists argued that trait theories could only help in the selection of leaders as only those people who had the requisite personal traits would be eligible for the leadership role. The behavioural approach, they contend, allows us to train anyone to become a leader through the encouragement of individuals to behave in ways that are required of leaders. This approach therefore allows the number of potential leaders to be expanded.

The Ohio and Michigan studies

The first major behavioural studies of leadership were undertaken at the Ohio State University and at The University of Michigan. At the Ohio State University a questionnaire was devised in an attempt to identify independent behavioural characteristics of leaders. Analysis of the questionnaire revealed two distinct types of behaviour: these were labelled 'consideration' and 'initiating structure'. Consideration describes behaviour that is oriented towards people and is concerned with subordinates' feelings and the showing of respect for subordinates' ideas. Initiating structure describes behaviour that is task-oriented and concerned with the structuring of subordinates' roles in order to attain goals established by the manager. Although the focus of the research was somewhat different, the University of Michigan studies produced similar results to those found at Ohio and also identified two main behavioural dimensions which they referred to as employee orientation (similar to 'consideration') and production orientation (similar to 'initiating structure').

At Ohio it was found that the subordinates of managers who scored highly in both categories achieved good performance and satisfaction more frequently than those working for a manager who scored low in one or both categories. It is, however, worth noting, that scoring highly in both categories does not guarantee good results from the workforce. For instance, scoring high in the 'initiating structure' and therefore demonstrating a strong task orientation can lead to a decrease in job satisfaction, an increase in grievances, absenteeism and staff turnover. At the same time, a strong people orientation ('consideration') may give the impression of weakness on the part of the manager and lead to a loss of confidence by the workforce. The balance (between the categories) required for success depends on the situation prevailing at the time. Examples here might include the imposition of task structure by virtue of the technology being used, time pressures, what the workforce expects and how compatible their style(s) are with those of their leader.

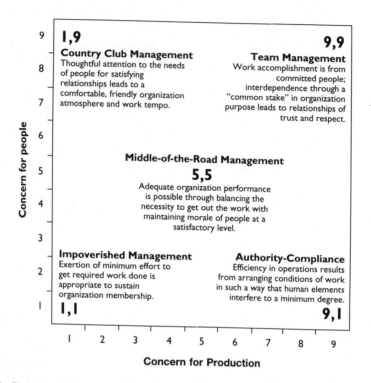

Fig 4.4 Blake and Adams McCanse's leadership grid. (*After* the Leadership Grid® Figure in *Leadership Dilemmas – Grid Solutions* by Robert R. Blake & Anne Adams McCanse (formerly the Managerial Grid figure by Robert R. Blake & Jane S. Mouton) published by Gulf Publishing Company, Houston. ©1991 Scientific Methods, Inc. Reproduced by permission of the owners.)

Blake and Adams McCanse's leadership grid

The concerns for people and production identified in these early studies form the basis of what today is probably the best known two-dimensional theory of leadership: Blake and Adams McCanse's leadership grid, Figure 4.4.

The grid forms a matrix of possible management styles and is based on similar dimensions to those employed by both the Ohio (consideration/initiating structure) and Michigan (employee/product orientation) studies. In Blake and Adams McCanse's grid these axes are shown as concerns, for people and production. The grid does not show results, but identifies what are considered to be the dominating factors of a leader's behaviour in respect to attaining results. From a possible total of eighty-one grid positions, five are here distinguished as being of key importance.

1,1 *Impoverished management.* Also known as laissez-faire leadership. Managers in this position have little concern for people or productivity, avoid taking sides and stay out of conflicts. They do just enough to get by.

1,9 *Country Club management.* Managers in this position have great concern for people and little for production. They avoid conflicts and concentrate on being well liked.

To them the task is less important than good interpersonal relations. Their goal is to keep people happy.

9,1 *Authority-Compliance*. Managers have great concern for production and little for people. They desire tight control in order to get tasks done efficiently. They consider creativity and human relations unnecessary.

5,5 *Middle-of-the-Road management*. Also known as Organization Man management. Managers in this position have medium concern for people and production. They attempt to balance concern for both but are not committed to either.

9,9 *Team management*. This leadership style is considered ideal. Such managers have great concern for people and production and work to motivate employees to reach the highest levels of accomplishment. They are flexible and responsive to change, and understand the need to change.

Blake and Adams McCanse conclude that the 9,9 (team) leadership style is the most effective for organizations. Whilst the grid may be good for conceiving of leadership styles, subsequent researchers (Larson *et al.* 1976) have pointed out that it does not provide us with any clues as to how we may characterize leaders. Indeed, whilst the 9,9 leadership style may be the most effective in certain situations, there would appear to be little evidence to support their conclusion that this style is the most effective in all situations.

This point can be demonstrated if we consider a simple example. Imagine a resident engineer being informed that the Readymix concrete lorries have arrived in readiness for the pouring of a structural slab in which the reinforcement fixers have only just finished their work. At this point a decision has to be made as to whether the reinforcement has been placed properly and the formwork has not been disturbed by the steel fixer's operations. This is no time for a discussion with other members of the construction team, an immediate decision is needed. In this situation we might quite reasonably expect the resident engineer to become completely task-oriented (i.e. utilizing a 9,1 style of leadership). When the slab is poured and the pressure is off, the selfsame engineer may well return to being more people-oriented in his approach. Whilst accepting that a crucial aspect of leadership is the ability to make last minute and immediate decisions, Black and Adams McCanse argue that in this case the resident engineer (through ongoing inquiry and critique) will be able to assume the responsibilities of team members. This, however, does not remove the likelihood of short term task orientation.

Contingency theories

What the above example demonstrates is a problem that none of the behavioural theories have been able to overcome. They have been unable to identify invariable linkages between examples of leadership behaviour and successful performance. This situation left theorists with a puzzle. If the complex phenomenon of leadership is not explained by identification of a number of isolated traits or by preferred behaviour, then how were they to clarify it? Clearly, the earlier ideas were all relevant on some occasions, but in reality which was relevant at which point in time depended on the situation faced at that time. There were and are not any universal situations. This led to the consideration of 'situational factors' and the rise of contingency theories.

House's path-goal theory

House (1971) and House and Mitchell (1974) studied how the behaviour of leaders affected the motivation of subordinates. They identified four particular types of leader behaviour:

Directive – here the leader gives firm guidance and clear instructions wherever possible. They organize the work of subordinates and makes sure that they are fully aware of what is expected from them.

Supportive – the leader is friendly and supportive towards subordinates, showing concern for their well being, needs and status.

Participative – the leader asks for subordinates' suggestions and takes them seriously into consideration when making decisions.

Achievement-oriented – the leader attempts to get subordinates to accept full responsibility for their work, sets challenging goals and expects subordinates to work as well as possible.

The Path-Goal theory rests on two propositions which are founded in McGregor's theory Y hypothesis:

1. If subordinates view leader behaviour as a source of satisfaction, such behaviour will be acceptable and satisfying to subordinates.
2. Leader behaviour will be motivational to the extent that such behaviour could help improve the working environment of subordinates by providing the guidance, training, support and rewards necessary for effective performance and such behaviour makes satisfaction of subordinates' needs contingent upon effective performance.

In short, the Path-Goal theory suggests that leaders must adopt a style that provides a path (through direction and support) which will lead subordinates to achieve their goals: furthermore, the leader must ensure that these goals are complementary to those of the organization.

Fiedler's model

Developed from research undertaken in the late 1960s, Fiedler's model of leadership considers that effective work groups depend upon a proper match being achieved between a leader's style of interacting with subordinates and the degree to which any particular situation faced by the group gives control and influence to the leader. Here we can identify two central and interdependent concerns: firstly that leadership is partly a function of personality and secondly that it is partly a function of situation.

Fiedler said that leaders can only be effective if their personality and style of leadership is well matched with a given set of situational variables. In other words, if the workers like the leader they will be easier to lead than if the leader is unpopular.

Based on a questionnaire designed to measure whether an individual is task- or relationship-oriented, Fiedler analysed personality on what he referred to as the 'least-preferred co-worker scale' (LPC). By task orientation he means a controlling, active and structured form of leadership whilst relationship orientation refers to a more passive, permissive and considerate form.

	I	II	III	IV	V	VI	VII	VIII
Leader–member relations	Good	Good	Good	Good	Poor	Poor	Poor	Poor
Task structure	Structured	Structured	Unstructured	Unstructured	Structured	Structured	Unstructured	Unstructured
Position power	Strong	Weak	Strong	Weak	Strong	Weak	Strong	Weak
Effective leadership style	Task-oriented	Task-oriented	Task-oriented	Relations-oriented	Relations-oriented	Relations-oriented	Either	Task-oriented

Fig 4.5 Fiedler's model of effective leadership. (*Adapted from* Fiedler, F. E., *A Theory of Leadership Effectiveness*)

Fiedler argued that there is a correlation between an individual's score in the LPC questionnaire and their style of leadership. A low score indicates a task-oriented style where the leader is unable to disregard any (negative) characteristics of a worker which are likely to hinder the accomplishment of the task. A leader who scores high in the LPC test is considered able to discount such features and maintain a strong interpersonal relationship with the worker, despite the effects of such traits on task achievement. This is referred to as a relationship-oriented style. The theory is based on the idea that it is too difficult to change an individual's leadership style and therefore assumes it to be fixed.

Once the leader's style has been identified it is necessary to consider the situational variables and to match the leader's style with them. Fiedler grouped what he saw as the key situational factors in determining effective leadership under three headings:

Leader-worker relations. To what extent do the workers trust, like and respect the leader?

Task-structure. To what extent are jobs or tasks defined and procedural?

Position power. How much power is conferred as a result of the manager's position? In particular this is related to hiring and firing, disciplinary matters, promotion of subordinates, salary increases etc.

According to Fiedler, the degree to which each of these three variables are in the manager's favour is the determining factor regarding effective leadership. Where all three are highly positive the leader's situation is most favourable. Conversely, where these factors are extremely negative, the leader's situation is most unfavourable. In these conditions a structured approach is considered to be best. If the factors are relatively neutral then a supportive approach is recommended.

Fiedler's model of effective leadership can be summarized as in Figure 4.5.

Critics of Fiedler's approach claim that it is both bound by methodology and limiting in the number of situational variables. Furthermore, they doubt the reliability of the LPC questionnaire and the quality of measurement of the situational variables. These points, they claim, call into question whether Fiedler's model can truly predict effective leadership. Bresnen *et al.* (1986) were particularly critical of Fiedler. Their study on leader orientation in UK construction site managers used

Fiedler's model to operationalize the construct of leadership orientations of site managers and performance across a specified range of situations. In the course of their research they intentionally added three situational variables, considered important in construction, to the three originally proposed by Fiedler. These were duration of contract, value of contract and proportion of directly-employed labour. Bresnen found that rather than Fiedler's three variables, there are two completely different controlling variables in terms of construction activity: the transient nature of project-based organizations on site and the phenomenon of subcontracting.

Despite these criticisms, the overall approach does provide a number of significant ideas for the evaluation of leadership effectiveness. In particular it draws attention to the need to consider situational factors.

The Hersey–Blanchard model

This is a contingency theory of leadership that emphasizes the maturity of the subordinates: maturity being defined as the ability and willingness of people to take responsibility for directing their own behaviour.

Central to this idea is that successful leadership is realized when a style of leadership is selected that is dependent (or contingent) upon the level of maturity of the subordinates.

As with Fiedler's model, Hersey and Blanchard use the two leadership behaviour dimensions of task and relationship. The main difference between the two models here is that Hersey and Blanchard go further by classifying each dimension as high or low and then combine them to form four leadership styles:

Telling – identified by a high task/low relationship style that features directive behaviour through the definition of roles and instruction of the worker in every aspect of the job.

Selling – where both task and relationship factors are high the leader is both directive and supportive.

Participating – a style that combines low task and high relationship approaches. Here the leader and subordinate share in decision-making with the leader's main role being one of enabling and communicating.

Delegating – a low task/low relationship style in which the leader provides little in the way of direction or support.

Combined with the behaviour of the leaders is the maturity of the subordinates. Again Hersey and Blanchard defined four stages which are illustrated in Figure 4.6 and are defined as:

M1 – the most immature level where workers are unwilling and unable to take responsibility for their work. These people are neither competent nor confident and require clear and specific directions as offered by a high task/low relationship approach.

M2 – motivation exists and is demonstrated through a willingness to undertake tasks, but the worker is unable to perform the work through a lack of necessary skills. This combination of low ability and psychological needs requires a high task and high relationship style of leadership.

M3 – motivational problems can be identified in this stage where people have the necessary ability for the job, but are unwilling to undertake the tasks required of them. Leadership here needs to be of a supportive, but non-directive kind.

M4 – the highest level of maturity. Subordinates require very little leadership in either a task or relationship manner. People in this group are both willing and able to undertake their work and to carry the responsibility attached to it.

Handy's best fit approach

Accepting that any preference as to leadership style is likely to be viewed subjectively, Handy (1985) developed a continuum of preferences and styles ranging from structured to supportive. This theory suggests that effective performance depends on how good a fit exists between the preferences of both leaders and subordinates and the demands of the task to be undertaken. Where a mismatch occurs between these components the likelihood exists that effectiveness will be impaired. In these circumstances, Handy suggests that the level of support offered to the leader by the organization will be a deciding factor in determining the outcome. Where this support is strong, the leader may well influence the subordinates to follow the leader's methods. Where support does not exist, or is weak, Handy believes (unlike Fiedler) that the leader may alter their own behaviour to suit.

Other contingency theories of leadership

Numerous other contingency theories of leadership have been devised in recent years which all show that appropriate leadership style to achieve effective performance is dependent upon situational factors and not simply upon

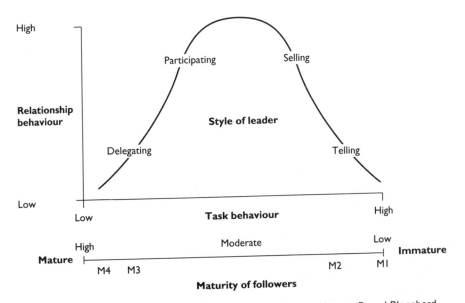

Fig 4.6 Hersey-Blanchard model of leadership. (*Adapted from* Hersey, P. and Blanchard, K., *Management of Organizational Behaviour: Utilizing Human Resources*)

either human traits or behavioural patterns. We might usefully touch on two of these.

Tannenbaum and Schmidt (1958) produced a graphical representation of the trade-off between a manager's authority and the freedom that subordinates have as the leadership style moves between being leader-centred and subordinate-centred.

Vroom and Yetton (1973) provide a means of determining the type and amount of participative decision-making acceptable in differing situations.

Fiedler's Cognitive Resource Theory (1987) extends his earlier work and introduces more factors than his earlier model in attempting to predict leadership effectiveness. Indeed, Cognitive Resource Theory could provide an alternative explanation for the discrepancies identified by Bresnen *et al.* mentioned earlier. Apart from LPC and the traditional contingency factors included in his earlier model, the new theory examines details relating to the leader, the group and the task. Examples might include the leader's experience and cognitive abilities (such as technical competence, intellectual ability and job-relevant knowledge), whether the task requires such cognitive abilities and so on. This new/extended theory suggests that the actual style used is, in part, determined by a combination of the LPC and control of the situational variables which will be dependent to some extent upon the leader's cognitive resources.

Motivation

Talbot (1976) found that labour represents approximately 40 per cent of total costs in a construction project. This cost will, of course, vary from project to project but clearly any reduction in labour costs will denote a direct financial saving to the organization. Levels of absenteeism, labour turnover, productivity, etc., are all commonly identified as directly attributable to employee motivation and so it is important for us to have an understanding of the factors likely to motivate (and dissatisfy) employees.

A problem for the manager of human resources is that two different concerns have to be reconciled here. Firstly, the requirements of the organization in terms of productivity and efficiency. These objectives are likely to result in people being viewed as instruments of production but who, through their efforts, are able to enjoy a satisfactory degree of prosperity and consumption. Secondly, the requirements of the worker in gaining satisfaction that is not directly associated with financial or material gain, but with enjoyment arising from undertaking the work itself.

Researchers have, over the years, attempted to find a link between these different aspects of motivation. Empirical study into the relationship between high productivity and high worker satisfaction remain, at best, inconclusive. Kahn and Katz (1953) and Blauner (1964) both concluded that the two might sometimes go together, but that this was certainly not an immutable fact. The uncertainty as to the relationship has not prevented managers from applying what might be considered to be a more humanistic style to their job, but as Thomason (1981) points out, 'it does face management with having to choose between alternative ends, or of establishing the priorities between them.'

Researchers have also proposed numerous theories that attempt to explain the nature of motivation at work. No generally acceptable theory has arisen and this may well be because the factors affecting motivation vary over time and according to circumstance and because of the sheer complexity of a subject that draws on a wide range of diverse psychological and sociological bases, e.g. anthropology. The complexity of the subject has led to theories being classified in numerous way. We shall consider the subject under two general headings: 'content' and 'process' theories.

Content theories

Content theories are concerned with what it is within an individual that generates behaviour, i.e. what is the specific nature of the driving force in an individual. The best-known of these theories include:

Herzberg's motivation/hygiene theory

Herzberg asked 200 engineers and accountants to recall incidents at work that had increased or reduced their satisfaction and what effects these incidents had on their attitudes and performance. He concluded that their responses could be grouped under two headings:

1. Hygiene factors which are primarily concerned with job environment and which are extrinsic to the job itself. These serve to prevent dissatisfaction.
2. Motivator factors which are mainly concerned with the content of the work itself. The strength of these affect satisfaction but not dissatisfaction.

A recent study of labour-only subcontractors in construction (Yap, 1992) has found that these definitive categories do not adequately represent the feelings of those currently employed in the industry and that the boundaries between them are much more vague than Herzberg proposed.

Maslow's hierarchy of needs

Maslow considered that humans have five identifiable needs (physiological, e.g. food and shelter; safety, e.g. security, protection from physical and emotional harm; social, e.g. affection, belonging, acceptance; esteem, e.g. self-respect, status, recognition; self-actualization, e.g. growth, self-fulfilment and achieving potential) and that these are ordered hierarchically according to whatever need is prepotent. Humans will not try to satisfy a need at the next level in the hierarchy until the lower-level need has been satisfied.

Maslow's work has been influential and has had a significant impact on management approaches to motivation and the design of organizations to meet employee needs. Whilst appearing instinctively logical, this theory has been widely criticized for lacking in empirical support. Several studies that have sought to validate it have been unable to do so.

McGregor's theory X and theory Y

McGregor believed that managers try to motivate people according to whether they hold a basically negative (X) or positive (Y) view of human beings.

Theory X was used to demonstrate what he considered to be the traditional approach to the direction and control of people. He claimed that this approach rests on the following assumptions:

- Workers have an inherent dislike of work and will avoid it whenever possible.
- Since workers dislike work, they must be coerced, controlled, directed and punished to achieve organizational goals.
- Most workers prefer to be directed, wish to avoid responsibility, have little ambition and desire security above all else.

Theory Y offers the opposite view of people and is based on assumptions that:

- Work is as natural as rest and play.
- Self-direction and self-control are desirable in the work situation.
- People are not inherently lazy. They become that way as a result of experience.
- People have potential. Under proper conditions they learn to accept and seek responsibility. They have imagination, ingenuity and creativity that can be applied to work.

McGregor's analysis corelates with Maslow's ideas in that theory X is akin to the lower order needs (physiological and safety) and theory Y to the higher order needs (social, esteem and self-actualization). McGregor believed in theory Y and felt that managers holding these views were more likely to successfully motivate their workforce than those who subscribed to theory X. Again, there is no evidence to support either of McGregor's theories.

Process theories

Process theories investigate how motivated behaviour is sustained. These more contemporary views are the result of empirical study and are generally given more credence by academics and those who have more faith in objective explanations than in the subjective 'gut-feeling'. We might do well here to give some thought as to whether it is indeed reasonable to insist on an objective approach to such a complex and human issue. Human beings tend not to be particularly objective in their feelings and behaviour, which might account for the continued acceptance of the early theories already mentioned.

Equity theory

This is concerned with the notion of 'fairness'. Developed by psychologists, this approach supposes that a major determinant of job satisfaction and performance is the individual's perception of the relationship between two variables: input and output. Input refers to the amount of work put in, whilst output is the reward (however that may be defined by the individual concerned) arising from the work. According to Adams (1965) and Pritchard (1969) people calculate this relationship in terms of a ratio, which they then compare with their own calculation of the same as it relates to others who they consider to be comparable in job terms.

In cases where the two ratios are perceived to be the same, equity is said to exist, i.e. the individual considers things to be fair. Where the perception is one of a difference between the ratios, inequity exists and the person will either consider themself to be under- or over-rewarded.

In cases of inequality, individuals are likely to find themselves in a state of tension and will be motivated to reduce it. Robbins (1991) suggests that individuals reduce or avoid it by:

- altering either their own or others' inputs and outcomes
- attempting to influence the other person to change their inputs or outcomes
- redesigning their job
- distorting either their own or others' inputs and outcomes
- comparing themselves with another individual

Earlier research by Landy and Trumbo (1976) found that people perceiving themselves to be overpaid will:

- reduce the quantity of production on piece rate, but increase the quality
- increase the quantity and the quality of production on time rates

Those, however, who consider themselves to be underpaid will, as they attempt to reduce the tension created by the inequality:

- increase the quantity of production on piece rate but decrease the quality
- decrease the quantity and the quality of production on time rates

Newcombe (1990) considers that equity theory offers an important implication for managerial practice particularly in designing jobs, reward systems and promotion policies.

Expectancy theory

This is derived from early work in the nineteen-thirties by Tolman (1932) and Lewin (1938) who laid the foundations for the work of Vroom (1964). Vroom established the idea of motivation being concerned with 'considered responses' to environmental stimuli. This work was later expanded on by Porter and Lawler (1968) into what they refer to as its 'expectancy–instrumentality–valence' theory or expectancy theory for short.

Expectancy theory assumes that individuals make decisions and behave according to what they believe the outcome of a particular action will be and how attractive that outcome is to them. This is what McCormick and Daniel (1985) refer to as a cognitive approach, based upon a rational-economic view of human nature. We might also consider it a theory based on self-interest where individuals know what they want, and assess the chances of getting it and the means by which it can be attained.

The core idea is that individuals perceive certain links between effort, performance and outcome. These are:

- effort–performance linkages
- performance–outcome linkages
- valence (attractiveness) of outcome

The relationship between these variables is shown in Laufer and Jenkins' 'expectancy motivation model' depicted in Figure 4.7.

Mescon *et al.* (1985) point out that if any of these links (critical to motivation) are weak, then motivation and subsequent performance is likely to be low. They express the relationship in terms of the formula:

$$\text{Motivation} = [\text{Effort--Performance}] \times [\text{Performance--Outcome}] \times [\text{Valence}]$$

Goal-setting theory

This is developed by Locke (1968) and stresses the significance of the setting and achieving of goals in the motivation process. The central tenet here is that an individual's conscious intentions (goals) regulate his/her action.

According to Locke, difficult goals lead to a higher level of output than easy goals. Furthermore, specific and difficult goals lead to a higher level of output than no goals or vague goals such as the simple requirement to 'do your best'. The belief is that people who participate in the setting of goals will work harder to achieve them than people who are simply assigned goals. The literature and research findings consistently support these ideas. As Hancock (1990) points out, 'We are all prepared to set tougher goals for ourselves than we will allow others to impose upon us'.

Mullins (1985) identifies the following practical implications for managers:

- Goal performance should be designed in order to direct behaviour and maintain motivation.
- Goals should be designed at a challenging, but realistic level.
- Feedback provides a means of checking progress on goal attainment and forms the basis for any revision of goals.

Goal-setting theory is conceptually aligned with the practice of 'management by objectives' (MBO). Although generally accepted as a good idea, the implementation of MBO has tended to fail in the past owing to the heavy layers of bureaucracy imposed on it and, because managers have been unaware of, or unskilled in the processes of agreeing goals, reinforcing them and provision of feedback. Hancock (1990) issues some cautionary words for when this style of management is practised on construction professionals: 'it is important to consider time and individual

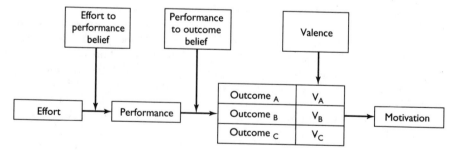

Fig 4.7 Expectancy motivation model. (*After* Laufer A. & Jenkins G. D. (1982) Motivating Construction Workers, in *Journal of the Construction Division, ASCE,* Vol. 108. Reproduced by permission of ACSE)

judgement, particularly when dealing with professionals who, after all, are being paid for their ability to make decisions on behalf of others and who are frequently involved in what we might describe as unprogrammable work. Be flexible, not slavish.'

With regard to the introduction of an MBO system he says that 'the gods, ghosts, sacred books and priests that control people's ideas cannot be defeated by logic. You will need to be willing to overcome the cultural patterns that oppose the introduction of change.'

McClelland's 'three needs' theory

A modern version of the needs-driven theories of Maslow and Herzberg, the McClelland's (1975) study of managerial staff concluded that there are three primary motives or needs in the work situation:

Achievement – the drive to excel, to achieve in relation to personally set standards of excellence.
Power – the need to control or influence, to make others behave in a way that they otherwise would not have done.
Affiliation – the desire for friendly and close interpersonal relationships.

McClelland noted that different people rank these needs differently. The ranking may be a result of individual personality, the result of one's position in the organization, or a combination of both.

It is interesting to note that those for whom achievement is a priority do not necessarily make good senior managers (although this need will be important in the early stages of a career). High achievers are generally more selfish and concerned with their own progress to be overly concerned with other people. At the same time good managers don't necessarily have a high need to achieve. Managerial success is frequently found in those who have a high power need (of a personal nature in junior and middle management and of an institutionalized kind in senior managers) and low affiliation needs.

Alderfer's ERG theory

Another needs theory, similar in type (although simpler) to Maslow, but supported by research, Alderfer (1972) identifies three main categories of need:

Existence – the need for survival and reproduction;
Relatedness – social needs and the need to be respected;
Growth – need for personal growth including the learning of new skills.

A major difference to Maslow's theory is that Alderfer believes that all three of these needs can be active at the same time.

Studies of motivation in construction

Although many studies of motivation have been undertaken, these have generally tended to be in manufacturing industries where the organizations studied have a permanent structure. The construction industry, with its temporary groups of project-based workers, has in the main been neglected and little recently published work exists. Why this should be the case is not entirely clear and different writers suggest different reasons for the omission. Both Schrader (1972) and Laufer and Jenkins (1982) believe that a lack of documented information is responsible. According to Maloney and McFillen (1983) there are three reasons why social scientists have avoided the subject in a construction context:

1. They have little or no knowledge of construction.
2. There is inadequate research funding from the construction industry.
3. Few construction researchers have a good understanding of the psychology and physiology necessary for such studies.

Not only has little research in the field been undertaken, but as Newcombe *et al.* (1990) noted many of the motivational 'tools' used by construction managers still arise directly from Taylor's principles of scientific management. This approach remains despite the numerous disadvantages of the classical school of thought and the fact that these have been identified and abandoned by the majority of other industries for some considerable time. The philosophy underpinning the Taylorist approach is a straightforward trade of rewards (or incentives) for performance. This may have been justified at the turn of the twentieth century, but there is doubt as to its current validity. Indeed, a study by Mackenzie and Harris (1984) shows that construction operatives are more likely to be motivated in a manner more in line with contemporary theories whilst the desire to merely increase wages is frequently replaced with higher-order needs such as job security, belonging and welfare.

Whatever the reasons for the lack of study and outmoded approach taken in practice, it is clear that more research does need to be undertaken if we are to improve our understanding of this aspect of human performance in construction. This idea is not new and was advocated back in the 1970s by Trench (1978) who suggested that construction management could improve the industry's effectiveness by giving consideration to the motivational potential of some of the ingredients in the work environment such as money, job satisfaction, a sense of belonging and a future. Although Trench's comments may appear rather simplistic, he does touch on a crucial matter, that is, the notion of the work environment.

Environmental factors

Theorists have recognized for some time that motivation is dependent upon the environment, the personality involved and the characteristics of the job or task. Basic needs may well be the same for all people in all industries, but the satisfaction of these needs will vary as a result of the controlling environment.

Environmental factors can be grouped under four main headings:

Economic factors:
- levels of competition
- levels of productivity growth
- degree of manufacturing
- service economy

Social/cultural factors:
- age composition of workforce

- level of education
- career expectations

Legal factors:
- employment legislation
- employment practices

Technological factors:
- computer-aided manufacturing
- degree of craft and/or knowledge-based work
- information systems and management

The above are only a brief selection of the variables within the construction environment. Many of these are significantly different from other industries and therefore specific studies of the motivation of construction operatives need to be undertaken if managers are to gain the maximum benefit from the techniques that they employ. From what has been said above, it seems clear that future research in this area should be undertaken either through a multi-disciplinary approach or by construction researchers with sufficient knowledge of the social sciences to be able to explain motivation in their industry.

Job design

As mentioned earlier, a difficulty facing the manager of human resources is in trying to motivate the employee whilst simultaneously achieving the objectives of the organization. Job design is an area where attempts are made to enrich the working experience of the employee in the belief that this will, in turn, increase the quality and/or quantity of that person's output. It is important here to avoid a crude notion of work simplification and time and motion studies, such as proposed by Taylor and the scientific managers. A much better approach is to consider the work experience in a holistic manner where the whole experience is considered rather than just the job or task itself. We can express job design in terms of four inter-related functions:

1. The process and principles employed by which the content of a job aimed at achieving the organization's objectives is determined.
2. The effective allocation of jobs among the organization's various employees.
3. Establishment of the relationships that should exist between the job holder and his/her colleagues, superiors and subordinates.
4. The establishment of the necessary personnel support system in terms of selection, training, supervision, assessment, rewards, etc.

The importance of a holistic approach to job design cannot be understated. If the work necessary to achieve the organization's objectives is not designed, allocated and arranged in an effective manner, then no amount of organizational structuring or improvements to the reward system will yield success. In short all of the points noted above need to be considered with no undue emphasis being placed on any one.

Early work in this area was undertaken in the 1940s by Trist and Bamforth at the Tavistock Institute. Their approach aimed to incorporate job design into the whole operating production system of the organization. This approach was then the basis for work undertaken in the 1960s by a Norwegian team led by Einor Thorsrud who experimented with the idea of autonomous work groups in which each group of workers held responsibility for planning and scheduling as well as doing the work. Group leaders were then responsible for liaising with other work groups and staff units.

An alternative to the socio-technical approach above is the job enrichment approach. This is based on Herzberg's motivator-hygiene model in which hygiene factors are seen to prevent dissatisfaction. The idea is that in order to motivate employees to better production it is necessary to enhance the quality of the work by making it more challenging and furnishing greater opportunities for personal advancement, achievement and recognition.

Two terms are commonly used with respect to this approach:

Job enrichment – the vertical restructuring of jobs designed to increase the level of challenge through more difficult duties and responsibilities. Whilst there is some evidence of improved quality through this method, improved production is harder to prove. Managers must also note that additional costs are likely to be incurred through the need for extra training.

Job enlargement – the horizontal restructuring of jobs whereby the degree of difficulty remains the same, but the number of duties is increased, providing more variety of work. The rise of the multi-skilled construction operative (particularly in the USA) may be held up as testimony to the success of this approach, although the traditional craft sectionalism prevalent in the UK construction industry might lead to a number of difficulties for construction managers here.

The particular approach taken to job design will be contingent upon situational factors faced by the organization and its managers at any given time, but we might usefully view the task as a three-stage process which can normally be applied.

Stage 1: Survey and analysis
Examine and identify:
- the content of a job.
- frequency of performance of specific tasks.
- degree of specialization in tasks
- importance of tasks relative to achievement of organizational goals

Consider situational factors, such as:
- economic position of and plans for the organization
- amount of technology employed
- working conditions
- changes in skill levels

Use interviews, surveys and tests to evaluate psychological and personal traits of employees.

Stage 2: Plan and design
Define systems and procedures required to realize organizational goals in the most efficient manner. This may be aided through the use of operations research or industrial engineering techniques.
Embody job operations in the newly defined system.

Stage 3: Implementation and assessment
Put the new design into practice (for a trial period if possible).
Assess and modify according to:
- compatibility with organizational goals
- feedback from employees

Review the process on a regular basis and encourage continued employee feedback in order to ensure:
- continued efficiency and effectiveness
- employee satisfaction

Review

Construction projects are complex and undertaken in a highly competitive environment. Construction organizations therefore face high levels of risk and uncertainty. If they are to survive they need to plan. The role of the human resources manager in organizational planning and design exists at two levels: the macro level concerned with overall structure and the micro level concerned with job design and motivation.

The continual interaction of organizations with their environment also requires them to adapt in order to survive. Adaptation often means change and firms need to be able to cope with and implement such change with the least possible disturbance to their productive activities. One means of dealing with this is through the implementation of Organizational Development programmes. Organizational Development systematically analyses and seeks to understand the organization's environment and culture in order to identify what changes are likely to arise and the problems that will then ensue. In doing this, Organizational Development aims to improve the way(s) in which organizations deal with change and in particular with the human aspects such as participation, communication, conflict and interaction.

Construction organizations can be structured as line and staff or matrix models. The project-based nature of the industry has given rise to a combination of the two which allows both the formality of the line and staff structure and the lateral, more relaxed communication channels of the matrix model to co-exist.

The success of an organization can be measured in terms of the degree of satisfaction that its activities provide for those with an interest in it. Interested parties include both the general society and employees, so organizations need to consider their human resources as well as their markets and products.

The project-based nature of construction with its temporary multi-organizations leads to problems of fragmentation. A result of this is the creation of a complex and turbulent internal environment that exists alongside the generally hostile external environment. Coping with this requires both good management and strong leadership. Studies of leadership have evolved from social psychology. Whilst much research has been done and numerous theories developed they can be broadly grouped into three categories; trait theories, behavioural theories and contingency theories.

Given the labour-intensive nature of building works any reduction in labour costs will mean a direct financial saving to the organization. Problems such as absenteeism, labour turnover and low productivity are all linked to employee motivation so human resources managers must have an appreciation of the factors likely to motivate (or dissatisfy) employees. A difficulty here is in reconciling the requirements of the organization, for productivity and efficiency, with the requirements of the employee, for job satisfaction. Research aimed at resolving this puzzle remains inconclusive. Similarly, no general theory of motivation has been developed. It is suspected that this is because the factors affecting motivation vary over time and according to circumstance.

Most of the studies of motivation undertaken have been in manufacturing industries characterized by permanent organization structures. Little published work exists that is directly concerned with construction. It is suspected that a lack of knowledge of the industry on the part of social scientists and a lack of understanding of social science by those in the industry has been the cause of this dearth.

An approach that is believed to be helpful in improving morale and motivation amongst those in the industry is through job design. It is anticipated that the general enrichment of the work experience will increase the quality and/or quantity of the individual's output. Two main approaches can be taken to enhance the quality of work. These are job enrichment, where more difficult duties and responsibilities result in more challenging work, and job enlargement, where the level of difficulty remains the same but the number and variety of duties is increased.

Questions

1. Discuss the arguments that might be made both for and against the employment of an external consultant to carry out the process of Organizational Development.
2. How might ideas about motivation influence appropriate leadership styles in construction firms?
3. Consider the advantages and disadvantages of different forms of organizational structure in respect to:
   (a) a construction firm
   (b) a construction project

Bibliography

Adams, J. S. (1965) Inequity in social exchange, in Berkowitz L. (ed), *Advances in Experimental Social Psychology*, **2,** Academic Press, pp. 267–300

Alderfer, C. P. (1972) *Existence, Relatedness and Growth*, The Free Press, New York

Argyris, C. (1965) *Organization and Innovation*, Irwin

Armstrong, M. (1988) *A Handbook of Personnel Management Practice* (3rd Edition), Kogan Page, London

Blake, R. R. and McCanse, A. A. (1991) *Leadership Dilemmas – Grid Solutions*, Gulf Publishing, Houston

Blauner, R. (1964) *Alienation and Freedom*, University of Chicago Press

Brech, E. F. L. (1975) *The Principles and Practice of Management*, Longmans, London

Bresnen, M. J., Ford, J. R., Bryman, A. E., Keil, E. T., Beardsworth, A. D. and Wray, K. (1986) Labour recruitment strategies and selection practices on construction sites,

Construction Management and Economics, **Vol 4,** pp. 37–55

Bryant, D. T. and Niehaus, R. J. (1977) *Manpower Planning & Organization Design*, Plenum Press, New York

Chruden, H. J. and Sherman, A. W. Jr. (1976) *Personnel Management* (5th Edition), South Western Publishing Co., Cincinnati

Crane, D. P. (1979) *Personnel: The Management of Human Resources* (2nd Edition), Wadsworth Publishing Co. Inc., Belmont, CA.

Crichton, A. (1968) *Personnel Management in Context*, Batsford

Desatnick, R. L. (1972) *Innovative Human Resource Management*, American Management Association

Emery, F. E. and Thorsrud, E. (1969) *Form and Content in Industrial Democracy*, Tavistock Publications, London

Etzioni, A. (1964) *Modern Organizations*, Prentice-Hall, Englewood Cliffs, N.J.

Fiedler, F. E. (1987) *A Theory of Leadership Effectiveness*, McGraw-Hill

Fiedler, F. E. and Garcia, J. E. (1987) *New Approaches to Effective Leadership*, Wiley, New York

Flamholtz, E. G., Randall, Y. and Sackman, S. (1986) *Future Directions of Human Resource Management*, Institute of Industrial Relations, University of California, Los Angeles

Foulkes, F. K. (ed.) (1986) *Strategic Human Resources Management*, Prentice-Hall, Englewood Cliffs

Fryer, B. (1990) *The Practice of Construction Management* (2nd Edition), Blackwell Scientific Publications, Oxford

Graham, H. T. (1983) *Human Resources Management* (4th Edition), MacDonald & Evans, Plymouth

Guest, D. (1984) What's New in Motivation?, *Personnel Management*, May, pp. 20–23

Hall, D. T. and Goodale, J. G. (1986) *Human Resource Management – Strategy, Design and Implementation*, Scott, Foresman & Co., Glenview

Hancock, M. (1990) Towards industrial democracy for construction managers, in *ICTAD Journal Vol. 2 No. 1*, pp. 90–93

Handy, C. B. (1985) *Understanding Organisations*, Penguin, London

Heisler, W. J., Jones, W. D. and Benham, P. O. Jr. (1988) *Managing Human Resources Issues – Confronting Challenges and Choosing Options*, Jossey–Bass, San Francisco

Hersy, P. and Blanchard, K. (1982) *Management of Organizational Behaviour: Utilizing Human Resources*, Prentice-Hall, Englewood Cliffs

Herzberg, F., Mausner, B. and Snyderman, B. (1959) *The Motivation to Work*, John Wiley & Sons, New York

House, R. J. (1971) A Path–Goal theory of leader effectiveness, in *Administrative Science Quarterly*, **16,** pp. 321–38

House, R. J. and Mitchell, T. R. (1974) Path-Goal theory of leadership, in *Journal of Contemporary Business*, **Vol. 5,** pp. 81–94.

Kahn, R. L. and Katz, D. (1953) Leadership practices in relation to productivity and morale, in Cartwright, D. and Zander, A. *Group Dynamics*, pp. 612–28

Katz, D. and Kahn, R. L. (1966) *The Social Psychology of Organizations*, Wiley

Larson, L. L., Hunt, J. G. and Osborn, R. H. (1976) The Great Hi-Hi Leader behaviour myth: a lesson from Occam's Razor, in *Academy of Management Journal*, December, pp. 628–41

Laufer, A. and Jenkins, G. D. Jr Motivating construction workers, in *Journal of the Construction Division, ASCE*, **108, No. CO4,** December, pp. 531–545

Lawrence, P. R. and Lorsch, J. W. (1967) *Organization and Environment*, Harvard

Locke, E. A. (1968) Towards a theory of task motivation and incentives, in *Organizational Behaviour and Human Performance*, **3,** pp. 157–189

Maloney, W. F. and McFillen, J. M. (1983) Research needs in construction worker performance, in *ASCE Journal of Construction Engineering and Management*, **109,** pp. 245–254

Martin, D. D. and Shell, R. L. (1988) *Management of Professionals – Insights for Maximising Co-operation*, Marcel Dekker Inc., New York

Maslow, A. H. (1954) *Motivation and Personality*, Harper & Row, New York

McClelland, D. C. (1975) *Power, The Inner Experience*, Irvington, New York

McCormick, E. J. and Daniel, I. (1985) *Industrial and Organizational Psychology* (8th Edition), Allen & Unwin, Australia

McGregor, D. (1960) *The Human Side of Enterprise*, Penguin

Megginson, L. C. (1967) *Personnel Management – A Human Resources Approach*, Richard D. Irwin, Inc., Homewood, Illinois

Megginson, L. C. (1968) *Human Resources: Cases & Concepts*, Harcourt, Brace & World Inc., New York

Mescon, M. H., Albert, M. and Khedouri, F. (1985) *Management: Individual and Organizational Effectiveness* (2nd Edition), Harper & Row, New York

Mintzberg, H. (1975) The manager's job: folklore & fact, in *Harvard Business Review*, July–August

Mullins, L. J. (1985) *Management and Organizational Behaviour* (2nd Edition), Pitman

Newcombe, R. (1990) *Management: The Micro Level*, MSc in Construction Management by distance learning: management principles module, University of Bath

Newcombe, R., Langford, D. A. and Fellows, R. F. (1990) *Construction Management: Organisation Systems Vol. I*, Mitchell

Peterson, R. B. and Lane, T. (1979) *Systematic Management of Human Resources*, Addison-Wesley Publishing Co., Reading, Mass

Peterson, R. O. (1980) Human resources development through work design, in Edwin L. Miller, Elmer H. Burack and Maryann Albrecht (eds), *Management of Human Resources*, Prentice-Hall, Englewood Cliffs: reproduced from *Training and Development Journal*, August 1976

Porter, L. W. and Lawler, E. E. (1968) *Managerial Attitudes and Performance*, Irwin-Dorsey

Pritchard, R. D. (1969) Equity theory: a review and critique, in *Organizational Behaviour and Human Performance*, **4,** pp. 176–211

Robbins, S. P. (1991) *Management: Concepts and Practices* (3rd Edition), Prentice-Hall, New Jersey

Rowlinson, S. and Tsang Man Cheung (1990) Leadership style in construction managers in Hong Kong, in *Proceedings of CIB 90, W55–65 joint symposia*, Sydney, March 1990, pp. 388–399

Salaman, G. (ed.) (1992) *Human Resource Strategies*, Sage Publications

Schrader, C. R. (1972) Motivation of construction craftsmen, in *ASCE, Journal of the Construction Division*, **98, No. CO2,** September, pp. 257–273

Talbot, P. J. (1976) Financial incentives – do they work? in *CIOB Occasional Paper 10*

Tannenbaum, R. and Schmidt, W. H. (1958) How to choose a leadership pattern, *Harvard Business Review*, March–April, pp. 95–102

Thomason, G. (1981) *A Textbook of Personnel Management* (4th Edition), Institute of Personnel Management, London

Trench, P. (1978) Construction: a vehicle for regeneration, in *Building Technology and Management*, July/August, pp. 1–9

Trist, E. and Bamforth, K. W. (1951) Some social and psychological consequences of the longwall method of coal getting, in *Human Relations,* **4,** pp. 3–38

Wahba, M. A. and Bridewell, L. G. (1976) Maslow reconsidered: a review of research on the need hierarchy theory, in *Organizational Behaviour and Human Performance,* **15,** pp. 212–240

Vroom, V. H. (1960) The effects of attitudes on perception of organizational goals, in *Human Relations,* **13,** pp. 229–240

Vroom, V. H. (1964) *Work and Motivation*, Wiley

Vroom, V. H. and Deci, E. L. (1970) *Management and Motivation*, Penguin

Yap, C. H. (1992) *Herzberg and the Motivation of Labour-Only Subcontractors*, unpublished MSc dissertation, University of Bath

Chapter 5

Industrial relations in construction

The theme of industrial relations in the construction industry is no longer at the forefront of the public debate as it was, say, 20 years ago. The years of Conservative government have changed the setting and language of industrial relations and the political and economic environment is now antagonistic to labour engaging in collective activity. Yet despite the low profile of conflict between labour and employer in the building industry there remains a structure for the regulation of industrial relations: employers still need some semblance of labour organizations to talk to at the site and labour still retains trade unions as its collective spokesman albeit that the density of membership is low and the principal trade unions are beset with financial difficulties and organizational schisms. In this chapter the remaining components of the industrial relations system will be discussed and a prognosis for the future will be presented. It is convenient to break the chapter down into four sub-themes:

1. The labour force – its composition, structure and characteristics.
2. The employers – their size and level of fragmentation.
3. The industrial relations system in construction.
4. The future direction of industrial relations.

The labour force

Construction as an industry has a poor record in being able to attract and hold high calibre labour. Gale (1991) has documented the social barriers to young people entering the building industry. The industry is perceived to provide an unattractive and uncertain career and whilst this may not be matched by reality it is true to say that this perception remains. Negative images associated with working in the building industry are contrasted with the view of its products – the buildings themselves – which have risen in esteem so that the quality of the built environment has become a central public debate in the later part of the 20th Century. Indeed the products of the building industry are seen as critical for the 'socio-economic development of every country and have political, historical and cultural significance (yet) the lack of self-esteem of the workforce is particularly striking'. (Ofori, 1990).

The lack of self-esteem noted by Ofori has an obvious bearing upon the quality of the labour force in the building industry. The UN (1985) also found that construction

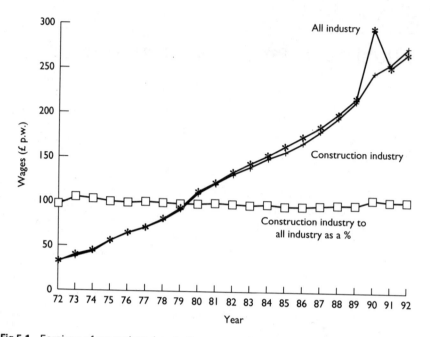

Fig 5.1 Earnings of manual workers in the construction industry and in all industries

work was associated with long hours, poor occupational health and safety and erratic provision of welfare services. Certainly data from the UK confirms the manual worker in building having to endure a longer working week for 14 of the years between 1972 and 1992. In all but four of these years wages in the building industry were lower than those for manual workers in all industry and services. Figure 5.1 shows the data extracted from Housing and Construction Statistics. Social attitudes provide further barriers to a high status labour force (although it is recognized that Japanese and Korean workers certainly do enjoy a prestigious position and unionized labour in the USA enjoy high wages if not high status).

Although attitudes are important, in reality the numbers of workers available and willing to work in the construction industry will be shaped by economic and social factors. Foremost will be the prevailing and prospective workload for the building industry; more subtly the mix of the workload will also shape demand for labour.

Figure 5.2 shows the trends for output and employment in the years 1972–1991. What is startling is the way the output and different types of employment are limited but after 1982 there is considerable distance between the total employed index and the output index. This may suggest a stronger informal sector than the Housing and Construction Statistics estimate since improvements in technology and greater use of prefabrication are not likely to close the gap between the output and employment indexes.

So much labour used in construction is unrecorded as much of the labour force is self-employed or employed as labour-only subcontractors. Langford (1985) argued that the 'informal' labour supply was in the region 50 per cent of the total. This estimate was based upon an amalgam of evidence drawn from the number of tax exemption certificates issued and Department of Employment data. Certainly the political environment through the 1980's favoured the growth of labour-only since the connection between the official wage rate and the market price for labour could only be loosely linked. This concept of a volatile labour market with workers 'pricing themselves into jobs' found favour with Government; a situation which was unrestricted by the intervention of trade unions or regulatory authorities such as the National Joint Council for the Building Industry. Figure 5.2 also shows the split between the recorded self-employed and contractors' operatives. In the 1980's the self-employed are shown to be supplanting directly employed workers as the main work force.

Subcontracted labour: the implications

It is unnecessary in this text to fully rehearse the traditional arguments for and against labour-only subcontractors (LOSC). This has been documented by Phelps-Brown (1968), Austrin (1980) and Langford (1985). Yet it is useful to refresh the key points. As late as 1987 the London Research Centre pointed out that the 'informal' economy created by the extensive use of LOSC tubulates the competitive bidding systems and distorts the labour market. The competitive bidding system is disrupted by a cash economy which invariably undercuts bona-fide contractors. The skills of the labour force are not developed since formal training is seldom carried out by LOSC. Yet those workers employed in bona-fide organizations are drawn to LOSC either for better wages or to work in a cash economy. To the employer LOSC has benefits beyond the obvious creation of labour flexibility. Most employers see the building industry as a means of generating cash. In economic cycles which favour investment this cash has frequently been sunk into property development or related diversification. The use of credit from material suppliers has enabled cash to be invested and the use of LOSC has extended the range of resources from which

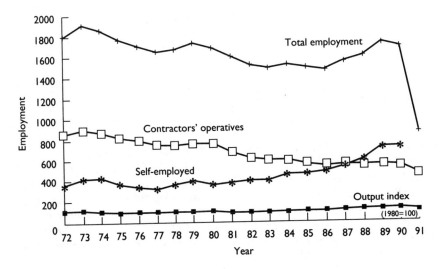

Fig 5.2 Output and employment

cash could be generated. Waterfall (1989) has noted that delayed payments, retention and tardy settlement of final accounts all create cash surpluses which may be used in investment opportunities.

With LOSC comprising such an important part of the labour force, both numerically and culturally, it was perhaps inevitable that one of the principal construction trade unions, UCATT, would seek a rapprochement with labour-only subcontractors. In 1988, after 20 years of hostility, UCATT conference rescinded the ban on recruiting labour-only subcontractors and this move enlarged the pool of eligible union members. Regrettably, as far as the trade unions are concerned, this liberalization of recruitment policy did little to boost membership since it coincided with the onset of recession. So, the separation of the bulk of the labour from direct control either by employers or organized labour has fragmented the labour force. In settings where labour is scarce this fragmentation enables labour-only workers to bargain up wages. Moreover, managerial control is frequently wrested from employers and production control is placed in the hands of a disparate labour force which is immune from the traditional sanctions which attend an employment contract. In short, employment conditions will be based upon various mechanisms of the market rather than a pattern governed by rules and procedures embodied in the National Working Rule Agreement.

The trade unions

The trade unions in the building industry are fragmented and whilst there has been considerable movement towards consolidation (the electricians and plumbers are now ensconced into the main engineering trade union) the principal union UCATT is frequently linked with possible mergers with many other unions including the Transport and General Workers Union (T&GWU) and the General, Municipal and Boilermakers Union (GMBU). Nonetheless, the unions in the building industry continue to have a presence in the industry. The principal actors are:

Union of Construction Allied Trade and Technicians (UCATT)
Transport and General Workers Union (T&GWU)
Amalgamated Union of Engineers, Electricians and Plumbers Trade Union (AUEEPTU)
Smaller unions involved in building include:
    General, Municipal and Boilermakers Union
    Furniture, Timber and Allied Trades Union
    Electrical and Plumbing Industries Union

Union of Construction Allied Trades and Technicians (UCATT)

The principal union in the building industry is UCATT. This union is by size in the 'top 10' trade unions, it claims a membership of 260,000 (Source: UCATT 1990) although this is challenged by the Labour Research Department (1992) who record a figure of 202,000 at the end of 1991. Of this membership approximately one-third are employed in the public sector. The union structure reflects the localized nature of the industry and has 12 regional offices which are staffed by 69 full-time officials. All officer posts are elected.

The structure of UCATT is shown in Figure 5.3.

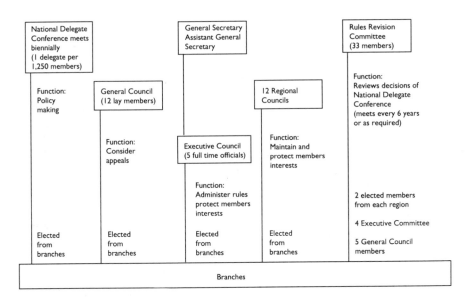

Fig 5.3 Structure of UCATT

Transport and General Workers Union (T&GWU)

The Transport and General Workers Union (T&GWU) is heavily involved in organizing building workers. It claims some 200,000 members in building-related activities. However much of the membership is drawn from those involved with the manufacture of building materials. Because the T&GWU is a general union, organizing workers in such diverse occupations as docks, manufacturing, transport, agriculture and building, the base unit of affiliation is the trade group. Each section within the union has its own national and regional offices. Fellows *et al.* (1983) describes the workings of the union thus: The grass-roots membership is organized into regional sections and, in general unions such as the T&GWU, workers in a particular industry are organized into industry-based branches such that each region will have building trade branches organized around a geographical area or a particular site. These branches elect one delegate to serve on the Regional Trade Group, which meets on a quarterly basis and has a full-time official to service it. However, quarterly meetings will not be able to cope with emergency matters and, consequently, an Emergency Trade Group is drawn from the Regional Trade Group. The Emergency Trade Group consists of three delegates from the Regional Trade Group plus a chairman and full-time official. Within the T&GWU there are eleven different trade groups, each of which elects a delegate to serve on the Regional Committee which overviews the work of the region. At the same time, the Building Trade Group will elect delegates to the National Committee of Building Trade Groups. This body holds meetings on a quarterly basis and its prime function is to review National Joint Council business. The National Committee of Building Trade Groups also elects one delegate to serve on the Executive Council of the union. The Executive

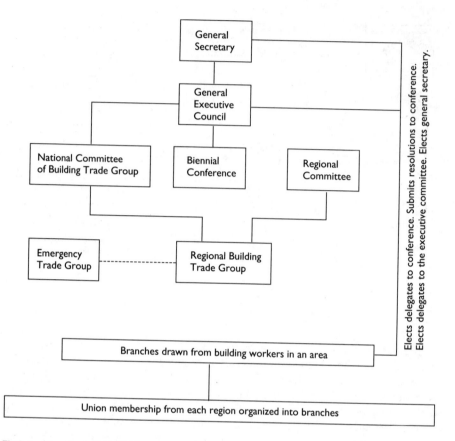

Fig 5.4 Structure of T&GWU

Council is served by the General Secretary of the Union. The union holds a biennual conference to which delegates are elected from the regions on a trade group basis.

The structure is shown in Figure 5.4.

General, Municipal, Boilermakers Union
This union is a party to the National Working Rule agreement. Its principal constituency are workers in local authority employment and has an estimated strength of 40,000 in the building trades.

Furniture, Timber and Allied Trades Union
Again this union is a part of the formal industrial relations structure. It concerns itself with organizing in joinery work and exhibition stand builders. It does not have a powerful presence in the industry.

The Electrical and Plumbing Industries Union

This union was formed in 1988 when the union organizing electricians and plumbers was expelled from the Trades Union Congress. Its remit was to organize electricians and plumbers into a TUC-affiliated union and it has a membership of approximately 5,000. Few employers recognize the union and it is destined to struggle for membership in the face of the recent merger between the engineers and the electricians.

The employers

A simple way of defining the nature of employers in the building industry is to consider the huge number of firms operating in the industry. The Housing and Construction Statistics records some 170,000 firms in operation. Of these, some 85,000 offer services as the main contractor, the remainder being specialists. Obviously most of these firms will be single person, sole traders and also many of the firms will be operating divisions of large concerns or conglomerates. Suffice it to say that there are a large number of small firms and a small number of large ones.

Using numbers of employees as a yardstick the pie charts shown in Figure 5.5 illustrate the division of output by the different sized firms. Snapshots of 1980 and 1990 have been taken to show changes in the industrial structure.

It is noticeable that the work done by the smallest firms has grown in the period 1980–1990. In some ways this picture obscures the industry structure since the larger firms do not seek to hold and retain labour but to subcontract work packages, with different degrees of continuity and allegiance, to the smaller firms. Figure 5.5 illustrates the way the smaller firms have grown in terms of providing employment and the larger firms which have shrunk over the period 1980–1990.

Figure 5.6 shows the relationship between output and total employment and contrasts the picture for 1980 and 1990.

Two points emerge from Figure 5.7. The first is that the larger the organization the less labour is required to undertake the recorded output. (Note that subcontractors will not be recorded as employed by the main contractors but equally their

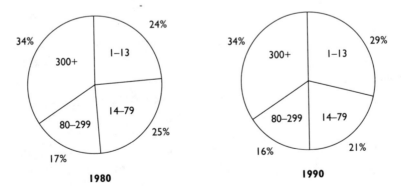

Fig 5.5 Proportion of total construction output by size of firm (number of employees)

output should not be recorded.) Secondly that the ratios have improved over the period 1980–1990 in each size category with the exception of the 1–13 group. However the improvements are again heavily slanted to the larger firms. The 14–79 group shows a 2 per cent improvement, 80–299 an 8 per cent benefit and 300+ improves by 16 per cent.

Whilst it is not possible to piece together a picture of these arrangements from the public information, it is self evident that the larger organizations are strongly differentiated from the smaller builders. The large organizations provide management services and the smaller ones undertake to build the physical product.

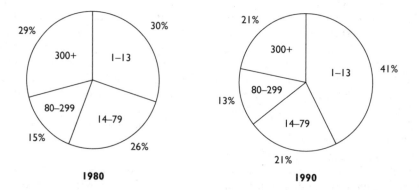

Fig 5.6 Proportion of total industry workforce by size of firm (number of employees)

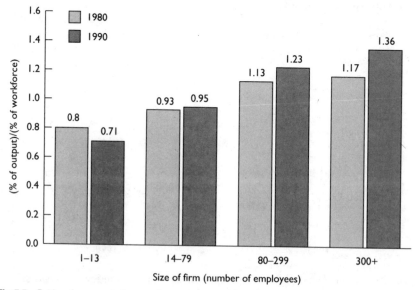

Size of firm (number of employees)

Fig 5.7 Ratio of per cent shares of output to workforce by size of firm

Further complications exist in interpreting statistics in that the largest contractors are highly diverse in their activities. Larger firms designated as building and civil engineering companies may be involved in many different market sectors. Typically the market will segment into:

Civil engineering
Building
Property development
Housing and materials extraction
Manufacture

The largest firms will be present in all market sectors whilst specialists may only be interested in a small slither of one of the market segments.

The principal employers associations are:

Building Employers Confederation (BEC)
Federation of Master Builders (FMB)
National Federation of Roofing Contractors (NFRC)
Federation of Civil Engineering Contractors (FCEC)

The nature of the industry and how its markets are divided has implications for the nature of employers associations. Firstly the relatively small number of large firms means that the principal employers federations are not numerically strong. This numerical weakness is compounded by the allegiance of employers to specialist trade or employers associations which are confederated into the BEC. Thus the National House Building Federation or the Federation of Specialist Sub-contractors or the National Contractors Group are part of the Building Employers Confederation but will claim the loyalty of members who have particular interest in the specialist areas.

This fragmentation has been recognized by building employers in the early 1990s. A co-ordinating body for employers was formed – the Construction Industry Employers Confederation. This body brought together the Principal Employers Federations, namely the BEC, FMB, the Federation of Civil Engineering Con-tractors (FCEC), the National Council of Building Materials Producers (BMP), the Federation of Associations of Specialists and Subcontractors (FASS) and the Federation of Specialist Contractors.

The Building Employers Confederation

This Confederation was founded in 1878 from an inaugural meeting of representa-tives of local associations of builders. Fellows, *et al.* (1983) have traced the development of employers associations and have linked a move towards national organizations of employers as a response to growing confidence and national organization amongst the trade unions. The confederation started life as the National Association of Master Builders in Britain but in 1901 it became the National Federation of Building Trade Employers. More recently, the name of the Building Employers Confederation was adopted in 1984. In 1994 it claimed over 5,000 firms in membership which represented 75 per cent of the building industries annual output.

The title 'Confederation' is appropriate since the BEC is organized into industry sectors. As of 1992 these sectors include *Building Contractors Federation* who constitute some 70 per cent of the membership and its role is to represent the interests of general builders.

House Builders Federation

HBF is solely concerned with private sector house buildings. All members of the HBF must be registered with the National House Builders Council – the consumer protection agency. It primarily acts as a spokesperson and pressure group for the speculative house building sector, making representation to local and central government and the agencies related to housing finance.

The National Contractors Group

NCG is open to the larger construction organizations. It is particularly concerned with industrial relations, contractual conditions, training, taxation and the public perception of the construction industry.

The Federation of Building Specialist Contractors

This association embraces some 100 different specialist interests and has in membership some 2,500 firms. The objectives of the federation are to promote the interests of specialist contractors and to harmonize relations between main contractors and their specialist subcontractors. Its principal interests relate to contractual matters, industrial relations and safety.

British Woodworking Federation

This Federation represents the interests of those engaged in the production of woodwork components for the building industry. Its work is focused on the provision of technical information but it has specialist committees dealing with industrial relations, marketing, training and safety.

Other trade associations

Each of the above federations is represented on the Standing Committees and the Council of the BEC. Linked to, but not within the Confederation are a number of specialist associations such as:

National Federation of Painting and Decorating Contractors
Federation of Plastering and Drywall Contractors
Stone Federation
National Association of Scaffolding Contractors

Local associations

The basic organizational unit of the BEC is the local associations. There are some 130 local associations throughout England and Wales which are attached to 10 regional centres. The structure is as shown in Figure 5.8.

The Federation of Master Builders

This organization claims some 20,000 small builders in its membership. Again the structure is based upon 126 local associations gathered into 10 regional councils.

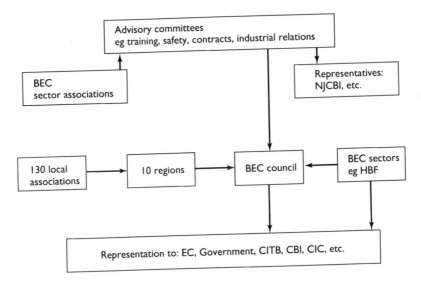

Fig 5.8 Structure of Building Employers Confederation

The work is principally making representation on behalf of smaller contractors. It participates (with the Transport and General Workers Union) in its own small industrial relations machinery in an organization known as Building and Allied Trades Joint Industrial Council (BATJIC). This body establishes wage rates and working conditions applicable to the smaller firms.

The role of the employers' associations

Thus the employers' organizations are numerically weak and internally subdivided. However, having recognized this, the employers' associations have several roles to perform on behalf of their members.

Economic

Employers have long since recognized that by combining they were more able to withstand demands for better wages and conditions from trade unions. It has been a long-established trade union negotiation tactic to 'leapfrog' in terms of wages and conditions, isolating one employer at a time. The employers' associations represent the employers' interests in a collective manner by undertaking negotiation with trade unions on a national basis over the question of wages and conditions of employment. The work carried out in this role is institutionalized in the annual agreement which amends the National Working Rule Agreement. This role is supplemented by giving advice and assistance to individual member firms when dealing with their particular labour problems. However this role has become diluted in the face of increasingly individual piecework negotiations with subcontractors.

The level of erosion of this function is evidenced by the difference between the minimum wage negotiated for craftsmen in 1990 (£148.20 per week including

Guaranteed Minimum Bonus) and the wages recorded by the Employment Gazette for manual workers in construction in the same year (£237). If one applies this to the average amount of overtime then the 'official wage' would increase to £178.20.

Advisory

Employers' associations provide members with an information service which is related to companies' trade or commercial functions. Areas included in this service would be impact of legislation upon the building industry, a wages monitoring service, research and development progress reports. In a sense, this particular service reflects the confederate structure of the building employers' association, where much of the lubrication for industrial relations is provided by specialist associations with the central body acting as a co-ordinator.

Regulatory

Employers' associations regulate and administer agreements they have reached on behalf of their members and generally provide facilities for the settlement of disputes between unions and individual managements. In this respect, the employers' associations attempt to stabilize relationships between the parties to the industrial relations machinery.

Representation

Employers' associations, in common with trade unions, seek to have their point of view made known to the decision-makers. In particular, it will make representations to government especially when seeking to amend specific legislation, but at a more general level, employers' associations will attempt to become party to economic planning in as much as any exists in respect of the building industry. Political support is also linked to the role.

Whilst the enthusiasm of Conservative politics amongst building employers may have waned in the recession starting in 1989 there remains a bedrock of support for the Conservatives. Many of the larger construction organizations recording donations to the Conservative party or its industrial arm – British United Industrialists – during the election year of 1989.

Technical and commercial services

The structure of the construction industry, with the largest number of firms being concentrated into the small to medium-sized organizations, means that an employers' association will often be asked to provide a technical and commercial advisory service. Issues which are likely to be foremost here are assistance in negotiating contractual conditions, legal advice, cost and estimating advice along with miscellaneous commercial advice for matters on which a small to medium-sized contractor would not have in-house expertise.

Industrial relations systems

Whilst acknowledging that the bulk of industrial relations is carried out at the place of work it is important to map out the work of the formal agencies dealing with industrial relations in building. At the heart of the formal industrial system is the National Joint Council for the Building Industry (NJCBI). This is a joint employer/union body with committees operating at National, Regional and Local level. Its role is to agree the National Working Rule Agreement (NWRA). This document sets down the minimum standards for employment conditions in the building industry.

The remit of the NJCBI is wide and this is reflected in the range of issues which the NWRA seeks to cover. The following list covers the areas of employment which are addressed in the document.

Wages
- basic rates
- extra payments for skill and responsibility
- bonus arrangements

Hours
- working hours
- overtime premiums
- shift work premiums
- night work premiums

Holidays
- annual and public holidays

Allowances
- travelling allowances
- arrangements for working away from home
- sickness and injury benefits
- tool and clothes allowances

Apprentices
- employment conditions
- training conditions

Safety
- hard hats
- safety representatives

Disputes
- disputes procedures
- termination of employment

Benefits
- death and retirement benefits

The issue of fixing the wages of building operatives is obviously one of the most important functions of the NJCBI, but the laying down of principles which govern conditions of employment is also a key function. However within the framework of rules there are regional and local conditions. In particular there are separate rules for Scotland and Liverpool and special travelling allowances for London.

Table 5.1 Working days lost through industrial stoppages 1972–1990

Year	Working days lost in:	
	Construction	All industries
1972	4,188	23,923
1973	176	7,145
1974	252	14,845
1975	247	5,914
1976	570	3,509
1977	297	10,378
1978	416	9,391
1979	834	29,051
1980	281	11,965
1981	86	4,244
1982	44	5,276
1983	68	3,754
1984	334	27,135
1985	50	6,402
1986	33	1,920
1987	22	3,546
1988	17	3,702
1989	128	4,128
1990	14	1,903

Supporting the agreement is a network of conciliation panels at local, regional and national levels. These panels are charged with resolving disputes which involve interpretation of the National Working rules.

In addition to the national agreement there are locally based procedural agreements used to manage industrial relations. These are site or company-based agreements which are primarily used to establish procedures for the settling of disputes concerning the interpretation of the NWRA. Consequently the remit of procedure agreements is likely to include:

- dismissal procedure
- redundancies
- recruitment and induction procedures
- trade union representation
- strike (or no strike) agreements, etc.

Disputes

Where disputes do occur the NWRA sets out the process by which the parties are to settle the dispute. The NWRA differentiates between 'individual cases', where operatives are urged to settle the matter with the immediate supervisor. Failing resolution at this level, the individual and the union representative can take the matter up with the site manager. Where the issue affects a group of operatives the matter is taken up with the shop steward or full time union official who should make representations to the site manager. If the matter cannot be resolved here then it proceeds through the local, regional and national conciliation panels.

Table 5.1 shows that the incidence of strikes has been relatively low. During the progress of a dispute the operatives are pledged not to strike or restrict production until the National Conciliation panel have met. Where a dispute involves the issue of wages, the Government's conciliation service, the Advisory, Conciliation and Arbitration Service (ACAS), is a service available to the employer and unions.

The future for industrial relations

The contours of industrial relations have changed remarkably over the period of 1972–1992. Quite clearly there has been an important shift in power during this period. If politics is about 'who gets what' (Lenin, 1919) and power is the instrument by which such decisions are made then it is self-evident that Government has sought to strengthen the political position of employers in respect to that of the unions. This opinion is reinforced by a retreat from collective activity – be this in the community or in the trade unions.

In the sphere of industrial relations the evidence of the move away from collective action is stark. Figure 5.9 charts the decline in the use of strikes over the period in question. (1972 has been used as a baseline as it was the year in which the first and only national building strike took place.)

But it is simplistic to see the temperature of industrial relations measured only by strikes. The reduction in the number of strikes may relate to changed tactics apart from the ability of strikes to shape events. Certainly the high unemployment experienced by the workforce in general and the building labour force in particular will have shaped perceptions of the likelihood of gaining advantage through strikes. The miners strike of 1986 will also have mellowed views about the use of the strike weapon. One may conjecture that the industry is seeing a revision in the way

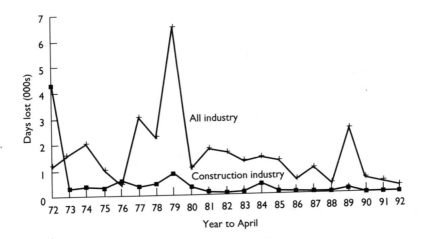

Fig 5.9 Working days lost in industrial stoppages (in 000s)

the labour force seeks influence at work. Industrial disruption and resistance to management control have become fragmented and individualized – two bywords of organizational behaviour throughout the 1980s.

Such industrial behaviour replaced earlier models of integrated and collective action but is likely to be equally disruptive to the work of the many trade contractors whose specialized labour is central to progress in modern building.

More positively, NEDO (1990) reported that craftsmen obtained considerable satisfaction from their work and had a strong sense of pride in their jobs. Yet the same set of workers feel driven by the inexorable rules of the market in that their piece-rates are driven down. In this setting the portents for harmonious industrial relations are not good, for there will remain the ever-present threat of conflict. How this conflict is expressed will surely be different from the 1970s and 1980s but nonetheless will be evident in the consequent effect on productivity.

It may be argued that the improvements in productivity obtained during a stable period of industrial relations has, in part, arisen from technical improvements in construction methods but more directly from the improvement of management methods. In particular management contractors or construction managers have used their expertise to improve productivity by creating strongly divided and specialized groups of labour. But this form of construction means that grievances are directed to subcontract employers but the action will frequently disable the whole site. This fracturing of the industrial relations system by using subcontractors means that the days when action was channelled into circumscribed forms (i.e. strikes, go slows, etc.) no longer exists. The end result may be sporadic, unstructured disputes with different groups demanding different concessions at different times necessitating a patchwork approach to the management of industrial relations. If such conditions come to pass the productivity gains of the 1980s may be tenuous and easily reversed and a strongly centralized industrial relations system may once again be seen as desirable!

Questions

1. In comparison with other industries, construction has a good record in industrial relations. Discuss the factors which gave rise to this situation.
2. Clients have a responsibility to improve industrial relations in construction. Discuss how the client can have a positive impact upon site industrial relations.
3. Are trade unions in construction governed democratically? Argue your case with reference to the structure and organization of the trade unions.

References

Austrin, T. (1980) The 'Lump' in the UK construction industry, in Nichols, T. (ed) *Capital and Labour: Studies in the Capitalist Labour Process*, Athlone Press

Faster Building for Commerce (1990) NEDO, HMSO

Fellows, R., Langford, D., Newcombe, R., Urry, S. (1983) *Construction Management in Practice*, Longman

Gale, A. (1991) What is good for women is good for men: theoretical foundations for action research aimed at the proportion of women in construction management, in *Practice Management: New Perspectives for the Construction Professionals*. Barrett, P. and Male, R. (eds.) Chapman & Hall, London, pp. 26, 34

Labour Research (1992) Labour Research Department, London

Langford, D. A. (1985) Labour-only sub-contractors, TIS paper 26, CIOB

London Research Centre (1987) *Skill Shortages in the London Building Industry*, report by the Economics Activities Group

Ofori, G. (1990) *The Construction Industry – Aspects of its Economics and Management*. Singapore University Press

Phelps–Brown, E. H. (1968) Report of the committee of engineering under Prof. E. H. Phelps-Brown into certain matters concerning labour in Building and Civil Engineering, HMSO

United Nations (1985) *Working Environment in the Construction Industry National Policies and Legislation in EC Countries*

Waterfall, M. (1989) The profitability of labour-only subcontractors, MSc thesis, Heriot-Watt University

Chapter 6

Interviewing for staff selection

In the construction industry the usual way of selecting staff for employment is by interviewing them, following the submission of an application form or CV. This chapter reviews the nature of interviewing and points out some of the shortcomings of interviewing in the context of the construction industry. It is recognized that interviewing is a strongly practical skill and pointers are made to company practice.

Interviews have often been criticized as an unsatisfactory and at times unfair method of selection, but, like the examination system, attempts to find workable alternatives have not been entirely successful.

An interview is the first meeting between the manager/employer and a possible new employee and whether we like it or not, it can set a positive or negative aspect to that person's period of employment and affect our future dealings with them. Therefore, selection interviewing requires careful consideration.

Selection interviewing and management practices

An interview has been defined as 'A meeting of two people, face-to-face, to accomplish a known purpose by discussion' (Brenner *et al* 1985). Consequently, interviews in an employment context fall within this definition with the exception of panel selection interviews which will be discussed later.

The definition is a powerful one and so it will be useful to anatomize it. If any reference to a 'purpose' is removed, what is left simply defines a conversation. Therefore, a conversation with a purpose becomes by definition an interview. (The corollary to this is that interviews can become conversations if the purpose is forgotten.)

The term 'known' is an important qualification of purpose. Interviews may not achieve what they were set up for because not enough was known in advance. This does not necessarily mean finding new information, but clarifying, organizing and understanding existing information. It also means that those doing the interviewing must, as far as is humanly possible, know themselves. The process of interviewing is surrounded by more myths, prejudices and half truths than many other management areas and training for this management process should ideally use theory supplemented by learning through experience and/or by the use of training aids such as videos, role playing and group discussions. The implementation of this is, however, hampered by the two most pervasive and dangerous myths about interviewing.

1. That all construction managers can interview just because they are managers.
2. In the case of selection interviewing, that 'gut feelings', 'hunches', 'I know the right people when I see them' attitudes are all that is required.

Bearing in mind that verbal information alone is not adequate to deal with this subject, if we return to our initial definition of an interview as a face-to-face conversation with a known purpose and add to it one of the definitions of management as 'getting things done effectively through other people', then it is theoretically and practically apparent that a large amount of management work will be done by interviewing, of one form or another, whether it is actually termed an interview or not.

Moving on through the definitions: 'Conversations with a known purpose' will be crucial to defining an interview, whether they mean:

*   selecting the right people to join the team
*   appraising them effectively
*   counselling them in difficulties
*   disciplining/reprimanding them if necessary

Human beings and human processes are fallible and construction managers often have little control over events that beset a construction organization. It can be argued that if every effort is made to select and recruit the right people for the managerial or professional jobs we have in mind, then there should be fewer problems later on with personnel disciplinary matters, personality clashes, labour turnover and other ills that fall upon the organization.

Selection interviewing

Knowledge, together with planning and preparation, are essential preliminaries to interviewing. In relation to selection interviewing, it is a matter of knowing what kind of person is being sought. This is partly a question of other effective personnel practices, especially manpower planning and comprehensive, accurate and up-to-date job descriptions. It may also be a matter of appraisals; if a vacancy has arisen, due either to departure or expansion, it is worthwhile bearing in mind that recruitment can be both costly and time consuming and may not find the right person at the end of it.

Knowledge of the work of a particular department, team or group and of the people in it will help the judgement of whether there is a suitable person for a promotion or transfer after a short period of training. Any subsequent recruitment may then be delayed or delegated elsewhere. Such an action may also have other benefits, including:

*   consolidating a team
*   enhancing morale
*   providing evidence that an organization values ability and performance and does not waste its talent

Clearly, a proportion of resignations are brought about by being passed over for promotion and then having a new boss introduced from outside.

Identifying recruitment needs

Whether you are recruiting from within or without, it is important to know what is required. Formal job descriptions can be so vague and/or out-of-date that they are useless. It could be beneficial both to construction organizations to generate their own by asking present and, if possible, previous job incumbents to write their own descriptions using a 'warts and all' approach! This means describing not only the satisfying, challenging and enjoyable parts of the job, but also the difficult, disagreeable and boring parts – bearing in mind that one person's satisfaction is another's boredom. This action should create a vision of the type of person being sought and clarify which gaps and weaknesses need remedying as well as the strengths which require enhancing. In an ideal world, this should be part of long-term planning.

To begin with, it can be helpful to list a small number of major responsibilities of the vacant position, i.e. those areas in which the new manager or professional will be spending the majority of time every day and then to list the crucial skills or special knowledge for each responsibility that a person must possess to carry it out.

Succession planning

In addition, if someone has left, it could be helpful, albeit difficult, to find out why, so as to avoid similar pitfalls in the future. In a crisis, the quickest solution may be to look for a similar person. This may not be the most effective long-term solution. Thus forward planning need not be altruistic; as well as looking for a good performance from a team, longer-term considerations will need to be considered. This will involve succession planning – selecting and training someone to move the organization forward.

Job qualities

In general, the planning and job description process should enable the preparation of a profile of the person you are seeking. There are a number of ways of analysing the qualities required, some of which are more theoretical than practical but they can be summarized as:

- ability to do the job
- willingness to do the job
- manageability when doing the job

More specifically if appointing, say, a manager then the organization may be looking for qualities which would include:

Major responsibilities:
- completion within budget
- completion within time constraints
- completion at appropriate level of quality
- developing good relations with clients and designers

Skills:
- technical skills and understanding of the relevant construction processes
- social skills related to managing very many different types of people
- diplomatic skills of handling problems and crises

Knowledge:
- education to BSc level with professional qualifications
- knowledge of the technical processes involved in similar projects
- knowledge of the safety codes and other relevant legislation

Ability, willingness and manageability

Ability to do a job can be assessed from a candidate's qualifications and experience. When looking at qualifications, it is important and necessary to distinguish between what educational/professional qualifications are necessary to do that job and what are desirable. When looking at experience, the firm deals with the additional proviso that quality may be more important than quantity. In other words, a shorter length of experience can possibly contain more depth and diversity than a longer time which may have involved the repetition of one or a few years experience.

Qualifications and experience are invariably included on an application form or curriculum vitae and are therefore relatively structured to deal with. Willingness and manageability are far less tangible and there is much confusion arising from words like 'character', 'behaviour' and 'personality'. These are value judgements and different jobs require different personal qualities. As with qualifications and experience, the future as well as the past must be considered. Too much similarity leads to stagnation; too much change, however, can lead to conflict and high turnover unless the organization is confident and clear-headed enough to embark upon change and find the people to implement it.

Quantifying abilities

In order to be more precise about these intangibles, it is useful to group characteristics in three profile categories:

- personality
- professional
- business

Some or all of the following characteristics could be important depending upon the particular job for which you are interviewing:

Personality:
- drive and ambition
- motivation
- communication skills
- leadership/co-operation skills
- energy
- determination
- confidence

Professional:
- reliability
- integrity
- dedication
- pride in work well done
- analytical skills
- listening skills

Business:
- efficient
- economical
- understanding and acceptance of procedures
- understanding of profit

The interview process

The next stage in the process leading up to the interview is finding the applicants. Whether this is done through advertisements, employment agencies, recruitment consultants or personal referrals, there is likely to be a tremendous amount of paper and paperwork, details and procedures. Whilst it may be desirable for senior staff to involve themselves in this process, it may not always be possible and some of it will have to be delegated to junior staff or to the Personnel department. Attention to detail at this stage is crucial because the whole matter of recruitment and selection is not a one-way process. Just as some applicants may not be acceptable to the firm, it is also possible that the organization may not be acceptable to the applicants; inefficiency and oversight of important details will not create a very good impression.

It is, however, likely that senior staff will wish to undertake the screening of application forms. The advice from practitioners is that in order to maintain objectivity and alertness do not attempt to review more than six in one sitting. This is also where the lists of major responsibilities prepared earlier come in useful. In comparing the job roles with the applications the following may be useful:

- Condense these lists by considering the most important facets of the job.
- Carefully consider the CV and write down the person's relevant qualifications and experience opposite the appropriate line.
- Begin to formulate the questions you may wish to ask should you call the applicant to interview.

Methods of interviewing

Interviewers have to operate in a number of different types of interviews. Each type of interview will have different characteristics and may be classified as:

- interview panels
- selection boards
- group selection

The interview panel

The interview panel consists of two or three people interviewing one candidate, the most usual combination includes a personnel manager and a site/line manager. The advantages are that information can be shared and the two interviews can discuss their first impression of the candidate's behaviour in the interview and modify any superficial judgements. The disadvantages are that the candidate may not be put at ease, there may be differences of view between the managers and it may be difficult to get all interviews together at the same place and same time.

Selection boards

Selection boards are more formal and are often convened by an official body because there are a number of parties interested in the selection decision. Institutions such as the Armed Services, Civil Service and Universities tend to favour this method. This may be due more to the innate conservatism of such institutions than to the advantage of this method, which is that the selection panel enables a number of different people to have a look at the applicants and compare notes on the spot. There are, however, a number of disadvantages which are that:

- The questions tend to be unplanned and delivered at random.
- The prejudices of a dominating member of the board can overwhelm the judgements of the other members.
- The candidates are unable to do justice to themselves because they are seldom allowed to expand and/or are intimidated.
- Boards tend to favour the confident and articulate candidate, but in doing so they may miss the underlying weaknesses of a superficially impressive individual.
- The board may underestimate the qualities of someone who happens to be less effective in a formidable and threatening selection, although he/she may perform well in a less intimidating interview and be able to do the job competently.

It is worth bearing in mind that these disadvantages do not apply to selection boards only. Unplanned and random questions, personal prejudices of the interviewer, not allowing the candidate to expand, being misled by a confident and articulate manner or the lack of one are all possible pitfalls in individual or panel interviews.

Group selection

Group selection is a time-consuming and expensive procedure and is usually only undertaken by large organizations and less conservative institutions. It involves gathering a number of candidates together (ideally six to eight) in the presence of a number of interviewers/observers (ideally two to three). The candidates go through a series of exercises and tests which are supplemented by the individual or panel interview. The exercises may be a case study which includes features and problems typical of the organization or they may be more general about a particular social, political or economic issue. The observers then rate or more usually rank the participants on such qualities as logic, practicality, leadership, confidence, listening skills and ability to accept criticism. In addition to the exercises and interview, candidates may be asked to present material either verbally or orally and can be given intelligence, personality or aptitude tests.

Just as selection boards illustrate some of the pitfalls of interviewing, group selection procedures highlight the purpose of interviewing which is to obtain the right people for the right position at the right time. Following from this is the question of validity, a concept borrowed from the psychology of assessment. There are several forms of validity; the two which concern us here are face validity and predictive validity.

A group selection uses face validity, i.e. it appears to measure what it sets out to measure because it exposes candidates to a number of more or less realistic situations and enables the interviewers and observers to see them in action individually and with others. It has also been found to have predictive validity, i.e. to predict future job performance. Those that did well and were selected after the group process were successful subsequently in the job.

However, the group selection is not the perfect answer to the interviewer's prayers. Vernon (1964), a pioneer in assessment psychology, gave the following warning:

> They are likely to be somewhat superior to the conventional interview method of assessing people because they provide a more prolonged and varied set of situations in which to observe and interpret. But they are just as dependent as the interview on the skill, expertise and impartiality of the observer and they should be applied with all the more caution because they engender in the observers an undue measure of confidence in the accuracy of their judgements.

This contains a timely warning to individual interviewers as well. A good interviewer also requires skills, experience and impartiality; over-confidence in the accuracy of one's judgements must be guarded against.

Prejudices

As stated earlier, one of the key skills of interviewers is self-knowledge and awareness of confidence in the interviewers judgement, intuition and assessment of people. Prejudice is an emotive word: a few people take a perverse pride in their prejudices. Most people, however, do not like to think that they are prejudiced and are reluctant to admit it to themselves, let alone to others. Realistic self-knowledge, however, reveals that most of us are prejudiced both for and against certain people or aspects of people. Interviewers who are aware of them can guard against being over-influenced by them.

Prejudice can be for certain aspects or against them. One 'for' prejudice that has always existed is also known as the 'halo effect', that is being overly-impressed by external and superficial aspects of a person. In earlier times, in the UK at least, social class, accent and appearance carried a lot of weight.

The opposite of the 'halo effect' is the 'cloven hoof effect', which is assuming from dirty shoes, a coarse accent, scruffy clothes and a weak smile, that you have someone idle, ignorant or unreliable. This may be off-putting, especially if the position requires a good appearance from positive first impressions, but it is better to withhold judgement until after the interview.

Other prejudices related to appearance which have been documented are against short people, fat people, too thin people, ugly people, bald men, bearded men, people with hobbies not considered appropriate to the gender (e.g. men who knit, women who weightlift), people who live alone regardless of marital status, people with a different sexual orientation and so on. With all of these, the prejudice says more about the interviewer than the candidate and the crucial question is 'would any of these factors affect the person's ability and willingness to do the job?'

The interview

We have now reached the point where ideally the arrangements have gone smoothly, the candidates have assembled and the interviewer(s) are prepared. Candidates and interviewers sit down together and are hopefully assimilating, without being unduly influenced by first impressions. Interviewers have their own preferences – and prejudices – about what should or should not happen next and what questions should or should not be asked. Given that all interviewers and interviewees are different there are few procedures which are absolutely right and must be followed at all times and few which are absolutely wrong and must consistently be avoided.

It will need to be borne in mind that an interview is a conversation with a purpose, not an interrogation or a question and answer session. A good interviewer should be able to 'draw out' the candidate to talk freely about him/herself and his/her career. However, it is all too easy, especially with confident and articulate participants, to be sidetracked into peripheral and irrelevant issues, so the interviewer must also plan, direct and control the conversation to achieve its purpose which is to make an accurate prediction of the candidate's future performance in the job for which he/she is being considered.

The interview approach

Most practised interviewers advocate a biographical approach starting with education and then moving through work experience, job by job, assessing each job, by asking questions such as:

- Why did you take the job?
- What did you do in that job?
- What knowledge and skills did you acquire?
- Why did you leave the job?

Most authors on the subject of interviewing agree that questions about what a candidate did in a previous job and the knowledge and skills that he or she acquired are equally important. However, there is some disagreement about the relative importance of questions dealing with why candidates take or leave jobs.

Those who emphasize the taking, and downplay the leaving, see it as being a positive, rather than a negative approach. They argue that too much emphasis on 'why did you leave?' can be seen as threatening by the candidate. On the other hand, those who do think that the 'why did you leave' question should be emphasized, regard it as a way of assessing not only honesty, but also other personality characteristics and as a way of dealing with the over-confident and glib candidate.

There are no hard and fast rules here, neither are there when it comes to the relative emphasis on education and job history. Usually with younger candidates, more time will be given to education: with older candidates more time will need to be spent on job history, especially the more recent jobs. Another difference of opinion arises with questions about interests outside work. Some writers regard these as a waste of time, others see them as useful. With younger candidates they are usually helpful, not so much with older ones.

One of the reasons for planning and preparation prior to the interview is to ensure that the information gathering does not go on for too long. It can seldom be completed in less than 20 minutes but is usually unproductive if extended beyond 30–40 minutes. Allowance has to be made for the information-giving part of the interview and for the candidate's questions. Interviews for management positions may take longer or require a second interview to discuss recent experience and ambitions more thoroughly. Another controversial area is whether to tell the candidate about the job, the company and discuss conditions of employment pay and fringe benefits before or after the biographical interview.

Those who say it should be done before, argue that any candidate who realizes that he/she and the job are not compatible can then leave without too much loss of face and time and energy are saved. Those who argue against say that the interviewer should establish the candidate's suitability first and the interviewer can then dispense with giving the information if the candidate is unsuitable and/or uninterested. A compromise is, probably, best in that general information about the job and the company can be given beforehand as part of the introductory, welcoming remarks and that more detailed information can be given after the interview according to what has happened during it.

Recording the interview

Analysis of candidate(s) against assessment criteria is again a matter of individual preference and a subject of some controversy. A pre-prepared checklist is preferable to taking notes during the interviews or being seen to do so. Listing experience, knowledge and skills and personality characteristics on one side with a space for comments on the other, can act as an *aide-mémoire* both during the interview and afterwards. Human memory being what it is, i.e. fallible and contrary just when it should be infallible and consistent, cannot be relied on! Therefore discreet recording during the interview and fuller recording afterwards is essential, especially when there are a number of candidates to interview.

Finishing the interview

Finishing the interview is probably more difficult than starting it: introductory remarks and small talk about travel, weather, etc. come more easily to most of us than saying goodbye! Whilst the interviewer may well feel a mixture of relief, regret and the need of a break, the candidate will want to know what the next step will be. One body of opinion argues that candidates should be told 'on the spot' whether or not they have been successful, others that candidates should be told that they will be hearing within a short space of time. People do not as a rule like to think that instant decisions have been made about them and there can be suspicions that the outcome was predetermined and the interview process a meaningless formality. The interviewers too need time to reflect on who they have seen and what they have heard and may need to discuss it with colleagues in other departments.

Dos and don'ts

As a summary here is a checklist for interviewing.

The dos of interviewing
- Plan the interview.
- Establish an easy and informal relationship.
- Encourage the candidate to talk.
- Cover the ground as planned.
- Probe where necessary.
- Analyse career and interests to reveal strengths, weaknesses, patterns of behaviour.
- Maintain control over the direction and time taken by the interview.

The don'ts of interviewing
- Do not start the interview unprepared.
- Do not plunge too quickly into demanding questions.
- Do not ask leading questions.
- Do not jump to conclusions on inadequate evidence.
- Do not pay too much attention to isolated strengths and weaknesses.
- Do not allow the candidate to gloss over important facts.
- Do not talk too much.

Questions

1. Identify the skills necessary to be a good interviewer.
2. Evaluate the methods of selection interviewing for the construction industry and suggest advantages and disadvantages for each method.
3. Your firm is recruiting staff. What major responsibilities, skills and knowledge would you attach to the job description for:
   - a site manager
   - a project architect

References

Brenner, M., Canter, D. and Brown, J. (1985) *The research interview – uses and approaches*, Academic Press
Vernon, P. E. (1964) *Personality Assessment – A Critical Survey*, Methuen

Chapter 7

Management development

Management development has been defined as (Langford and Newcombe, 1992):

> The process whereby the (construction) organization's managerial resources are nurtured to meet the present and future needs of the organization. This process involves the interaction of the needs of the organization and the needs of the individual manager in terms of development and advancement.

This may seem an idealized statement bearing little relationship to the reality of the construction industry but homogeneity, stability and cohesion within the industry and its component organizations seem to be the prerequisites for an effective and fruitful programme of managerial development. The construction industry, however, is rather unique and different when compared to other industries; it is heterogeneous and fragmented with mercurial qualities. Trying to pin it down in order to examine and define it is akin to stabilizing and confining mercury on a laboratory bench – maverick particles persist and insist on escaping in all directions.

It is not surprising therefore that in such an industry the working life of construction managers is characterized by variety, brevity and fragmentation brought about by the labour-intensive nature of the work which primarily involves organizing diverse and varied people to integrate their activities within a project setting. Fryer (1979) has said that co-ordination of people's activities and social skills are essential elements in a construction manager's job.

Effective management development must concern itself with the reality of the construction manager's job and management development involves the interaction of the needs of the organization and the individual manager. In order for it to be effective and fruitful, both the organization and the manager must agree on the reality.

However, there is not one single reality in the construction industry: there are several, including:

- the media reality
- the traditional reality

The media reality is fostered by television, newspapers and the advertising/public relations industries generally. The clean-cut machismo image, whilst very compelling, is more fiction than fact.

The traditional reality represents how the construction industry used to and in some cases still does see itself. This is a combination of media reality compounded with a pioneering, maverick and challenging image. Those working in the industry with this view see themselves as positively different from those in more traditional industries and regard working in an office and/or factory environment as akin to an imprisonment, albeit in security and comfort. They would never trade this for the freedom and challenge of working 'outdoors' and being very much their own boss. Consequently, traditional thinking about management development in construction has tended to hold a strong belief that 'only experience develops construction managers'. This belief however begs two questions:

1. What kind of experience develops the manager; what is to be the nature, quality and duration of that experience?
2. What are construction managers, and what are they supposed to do?

Langford and Newcombe (1992) state in reply that this view may have been valid in more stable times for the industry, i.e. when, strange as it may now seem, there was in the UK full employment, proper apprenticeship schemes, economic prosperity sufficient to fund building and civil engineering work and a continuity in building design, materials and methods.

However, it became apparent during the 1980s and early 1990s that this state of affairs no longer applied. The industry had never been as coherent and stable as other industries and the impact of so many changes in a relatively short space of time had shaken the industry to its foundations.

However, there are some certainties. The labour-intensive nature of the industry has remained constant. The economic, political and other changes in the industry, and the demands they make, have to be dealt with by the construction managers. Therefore it is imperative that a more structured, but flexible, approach is taken to management development in the construction industry so that its most precious resource can be used for the best for all concerned.

What is management development?

In 1969, The British Institute of Management defined management development as follows:

> The process of finding, keeping and developing managers to meet the current and future needs of the organization . . . and ensuring that the right people are developed in the right way.

In 1971 Morris used the definition:

> The systematic improvement of managerial effectiveness within the organization, assessed by its contribution to organizational effectiveness.

And in 1975 Ashton and Easterby-Smith had developed the definition into:

> A conscious and systematic decision action process to control the development of managerial resources in the organization . . . for the achievement of organizational goals and strategies.

Ashton *et al.* then elaborate on their definition by developing four perspectives involving both structured and conceptual (process) perspectives:

1. Management development can be seen as an organizational *function*, an integral part of the organization's structure, with characteristics similar to those exhibited by other functions such as personnel, finance and marketing.
2. Management development can be seen as a *style or philosophy* of management adapted by the organization. Here management development is seen as a reflection of the way things are done in the organization. The relationship and behaviour of managers in all aspects of work are constrained and guided by the organizational culture.
3. Management development can be seen as *bringing about change* in the manager's behaviour in order to achieve set organizational objectives. A good example is when management development is used to develop a sense of belonging and identity with the organization.
4. Management development can be seen as the *progressive development* of an individual manager's abilities at all stages of their career.

The structural perspectives, therefore, are concerned with the function and style of management, whereas the process perspectives are more concerned with modifying and developing a manager's attitudes and abilities.

Based on these perspectives the following important features of management development emerge:

- Management development relates to the overall organizational strategy, i.e. it should mirror and be congruent with the organization's present and future policies and strategy (Buckley and Kemp, 1987).
- The success of management development is measured by the extreme to which the organizational goals are achieved (Ashton and Easterby-Smith, 1979).
- Management development tries to harmonize the development of a manager, so far as is possible both in career and personal terms within the organization context (Leggatt, 1972).
- Management development is an integral part of the structure and culture of the organization.
- Effective management development is essential to the present and future effectiveness of the organization.
- The main thrust of these statements is that whilst some acknowledgement is made of the importance of management development to the development of the individual manager, the principal objective of development is to create people who will fit into the organization, i.e. in the image of the organization.

Management development, therefore, must be integral, effective and meet the company image. However, the organizations and the social and economic climate in which managers now find themselves are not static but dynamic. If it is to avoid stagnation, the organization must regard management development as future oriented, i.e. it cannot continually develop people solely in its present image but must take account of the risks of losing its most able and innovative managers if it does not take their needs into account during the planning and execution of its management development programmes.

What management development is not

Having talked about what management development is, we also need to consider what it is not.

> Management development is not limited to the formal activities of education and training but includes conscious and unconscious learning, informal – accidental learning processes and integrated – opportunistic learning processes (Mumford, 1987).

Langford and Newcombe (1992) point out that management development functions within the organization should not be confused with education and training.

The Manpower Services Commission (1981) defines education and training as follows:

> Education is any activity which is aimed at developing knowledge, skills, moral values and understanding of all aspects of life . . . the purpose of education is to provide conditions essential for people to develop an understanding of . . . the society in which they live and enable them to make contributions to it. Training on the other hand is a planned process to modify attitude, knowledge or skill behaviour . . . to achieve effective performance in an activity or range of activities. Its purpose is to develop the abilities of the individual and to satisfy the current and future manpower needs of the organization.

Education, then, has a 'soft' focus with its objectives being non-specific whereas training has a 'hard' focus and is concerned with more immediate needs. It is therefore possible to see education as dealing with the needs of the individual and training as answering the needs of the organization, but although management development partakes of both it is more than the sum of these two parts. To refer back to Mumford's definition: it is a personal as well as an organizational process involving growth or realization of a person's ability through conscious or unconscious learning.

Figure 7.1 illustrates the process.

In the context of construction management:

- Education is instruction in basic subjects, disciplines, techniques and principles relevant to the practice of construction management.
- Training is instruction concerned with developing skills required to practise specific managerial duties, e.g. to prepare a contract programme.
- Development is the search to integrate personal needs with those of the organization.

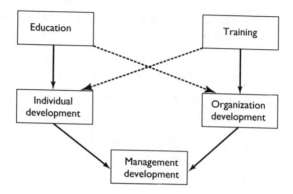

Fig 7.1 Management development process

Who should be developed?

The question of who should be developed is difficult and not easily answered. To begin to answer it, one needs to know the individual employees, their qualifications, skills, attitudes, experience, temperament, personality and personal circumstances. This information (which is not always collected or stored in any systematic way in the construction industry) needs to be balanced with the needs of the employer, both present and future. Like the systematic collection and storage of personal details, the construction industry has not always given consideration of these matters enough time and effort: because of the focus on activity and accomplishment, such considerations are often described as time-wasting navel-gazing.

Langford and Newcombe (op. cit.) draw on some of the available employment statistics in considering one way of answering the question. They say that the numbers of Administrative, Professional, Technical and Clerical (APTC) staff employed in the industry may be used as one indicator of the growth in demand for people who may benefit from some form of management development.

One reason for choosing this indicator has been the growth during the late 1980s and the anticipated continuation into the 1990s of the number of APTC staff. This growth has coincided with a shortage of such staff but it is important to remember that these statistics do not state an increase in absolute numbers, rather they indicate a decline in response to the 1978–83 recession in the UK through the proportion of such staff increasing.

Table 7.1 from the Housing and Construction Statistics (1989) charts the growth in the percentage of the APTC grade of staff.

In order to understand matters more fully, let us compare table 7.1 with table 7.2 which shows the shortage of management staff if the number of APTC staff are indexed and compared against the workload since 1980.

These two tables show that there was greater managerial intensity in the construction industry during the late 1980s – in other words, more managerial

Table 7.1 Numbers of APTC staff and operatives 1980–1988

Year	Workforce (in 000s)		APTC as a percentage of operatives
	APTC	Operatives	
1980	235	760	31
1981	236	659	34
1982	214	587	36
1983	214	600	36
1984	214	585	37
1985	213	586	36
1986	212	530	40
1987	220	541	41
1988	236	566	42

(*Source: The Housing and Construction Statistics 1989*)

Table 7.2 Index of APTC staff and workload 1980–1988

Year	Index (1980 = 100) of:	
	APTC staff	Workload
1980	100	100
1981	100	90.5
1982	91	91.8
1983	91	95.7
1984	91	99
1985	90	100.1
1986	90	102.8
1987	93	111
1988	100	129

(*Source: The Housing and Construction Statistics 1989*)

and technical staff were required to service the declining operative population. Langford and Newcombe (1992) state the following:

(a) The increased use of subcontractors on construction sites created demands for more co-ordination.
(b) Greater technical sophistication within buildings, e.g. the structural and services demands of information technology, have required more specialists and support staff who necessitate more management and co-ordination.
(c) Different procurement methods which have become more legally and financially complex have had the same effect as in (b) above.

Typical areas of management development

What are the areas that are the subject of management development? Some possibilities are:

- Environmental concerns, e.g. the necessary services to prevent toxic and noxious substances damaging the ecology of land, air and water;
- The changing, development and updating of our congested and dangerous transport systems.
- The need for both construction and economic innovation to restore our infrastructure of housing, schools, hospitals and sewage/drainage systems.
- Growing awareness on the part of companies that managers should be developed and encouraged to participate in the development process which is not altruism since coping with all the changes and economic pressures requires an ongoing improvement in managerial performance.
- Motivation. When people are given support and encouragement as well as independence in their work, and they are able to feel that their contribution to the organization is valued, then they are more likely to stay with the organization. Also, if an organization develops a reputation for good and effective management development, it is more likely to attract able and talented people at the recruitment stage, as well as obtaining optimal managerial performance from them afterwards.

The advantages of management development to the organization might include issues such as:

- public accountability – credibility could be increased if public sector organizations were seen to be using public funds wisely
- an openness to changing circumstances in a changing world

Advantages to the individual might include:

- increased self-esteem and confidence from improved job competence
- self-perception as an important part of the organization in which the individual works

Manpower audits

The benefits to the individual and the employer help organizations to identify management potential and indicate who should be proposed for management development. This process is called a manpower audit. To encourage this, the Construction Industry Training Board (CITB) has made grants available. For example in 1987/88 the CITB gave over £6m to aid management, supervisory and technical training in the UK. However, most commentators on the industry have argued that this is by no means enough, and a far higher investment is required by companies to sustain competence in a highly competitive market.

Williams (1987), in a review of management training for civil engineers, found a conflict between the recognition of a development needs and the failure to provide the resources to meet them. Employers, whether local authorities, consultants or contractors, were critical of the amount and direction of management development. The consultants and contractors particularly felt that junior staff should be developed in fields of leadership, communication, problem-solving and decision-making. Furthermore, they understood that construction organizations could do more to rectify the inadequacies that they perceived and that effective management development should begin at undergraduate/college level.

Williams found the answer was that employers were only providing development

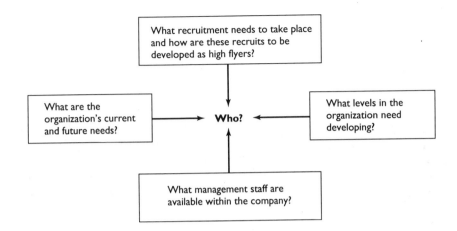

Fig 7.2 Decision process for development

opportunities for selected senior staff. One of the reasons for this presumably is that senior staff are more likely to be based at a head office and are not seen as having to be present 'at the coal-face', that is on site, virtually continuously to ensure project completion within time and cost constraints. In other words, they (senior staff) have the time and flexibility to attend courses, country house weekends, etc.

Williams considered the following issues important for the development of present and especially future managers in construction organizations:

- communication skills
- leadership
- problem-solving and decision-making
- organizational functions and objectives
- self-development
- task groups and motivation
- design technology
- personnel
- new materials
- civil engineering law
- computing
- politics and general legislation

Williams found that those organizations which were surveyed saw the main reasons for implementing development programmes as preparing their staff for future responsibilities and to satisfy organizational needs.

The benefits to the organization

There are two types of benefit of management development to an organization:

1. Those which accrue to the organization as a result of management development.
2. Those which stem from the results of training.

Results of management development

The organization which has implemented a management development programme enjoys the following advantages.

(a) If the training is effective, the people involved will be more efficient at their jobs. What is important here is that the criteria for effective training are understood; if not, the training will not be helpful.

(b) If the firm carries out an audit of its strengths and weaknesses, it will reveal areas where there are skill shortages impeding the efficiency of current operations. These gaps can be filled by a short-term management development programme.

(c) If the company is committed to developing future managers, then it will have to give serious attention to forecasting the future direction of the company so that gaps can be identified and filled by long-term development programmes.

(d) Following from (a) to (c) above, the management development programme can be used as a means of testing aspiring people to sort out the 'high-flyers' from the rest. Care needs to be taken in this because 'too many commanders and not enough infantry' could be harmful both to the individuals and to the company. The high-flyers can only succeed if they have the back-up of effective 'junior ranks' who should be valued just as much by the organization.

(e) An effective management development programme will make the company a more attractive prospect for able recruits – and a really progressive company will not only seek these among young graduates/college leavers, but also amongst older and more experienced people who, voluntarily or not, may be seeking a change of career direction and can bring experience and maturity to bear, both to the benefit of the company and to the younger recruits as potential mentors.

Results of training

The actual process of training managers also benefits an organization. A construction organization, to be successful, needs to complete projects within contractual time and cost limits. Managers with appropriate and up-to-date skills and training are more likely to be able to achieve this than those without. Training will result in:

- The ability to organize and integrate the varied work activity of diverse people within the setting and limits of a project.
- The ability to deal with the technical/environmental developments within the industry as well as with the changing legal, financial and safety rules and regulations.

In times of labour scarcity a company needs to be able to recruit able people, but it also has to retain them in a highly competitive market. People who change their job because of dissatisfaction with a present employer may do so for a higher salary and more fringe benefits but this will seldom be the only consideration. They will also be looking for a supportive, long-term management development programme which will indicate the value placed on the individual by the organization.

The benefits to the manager

A programme of self-development can be very motivating to ambitious people, and enhance confidence in those who are unsure. Many of the professional institutions in construction and engineering now insist upon mandatory continuing professional development (CPD).

Continuing professional development

Incorporation of this into a management development programme would enhance motivation and self-esteem among managers and also benefit the organization, especially in the provision of frameworks and guide-lines for the company if it was at all uncertain about how to proceed.

- Managers would be more effective and efficient and less prone to stress if they were able to keep up to date with the number and rapidity of changes.
- External courses can provide opportunities for managers to broaden their perceptions of their job through meeting people from other organizations. Discussing how problems have been solved, difficulties overcome, airing successes (and failures) and sharing common experiences as well as different ones; all provide a rich source of information and support in addition to the formal input of the course.
- Internal, in-house courses can also provide some of these experiences and, like external courses, provide a break from the daily/weekly routine which should recharge energy and provide different perspectives.

The practice of management development

The style and context of management development will take different forms in different firms within the industry. Consultants have one set of management development needs, contractors have another and those of local authorities will be different again. Also, within each firm the various professions and disciplines will have differing management development needs. Ashton *et al.* (1975) have suggested the following classification of management development in three phases which is portrayed in Figure 7.3

Phase I
This is characterized by a very low level of commitment by top management to development which is not perceived as part of the culture of the firm. There is only a small amount of development activity within the firm which is restricted to selected personnel being sent on external courses. This is reminiscent of the findings of Williams (1987).

Phase II
This suggests a higher level of commitment to management development accompanied by an increase in visiting activities. Examples of these are:

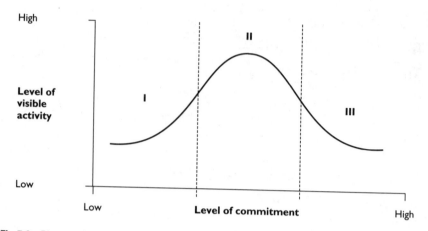

Fig 7.3 Phases of management development

- training courses become available within the firm
- training budgets to send people on external courses are expanded
- formal appraisal schemes have been developed
- audits of management talent within the firm have been undertaken

However these activities do not occur without conflicts about the orientation of the development activities which may focus around the two aims of management development, namely the achievement of organizational objectives on the one hand and the development of the individual managers on the other. It is possible that top management may attach a higher priority to extending the technical skills of staff so that the firm may enter a new area of the construction market whereas the managers themselves may see the development opportunity in terms of meeting individual managerial, professional and personal needs.

Phase III

During this, the level of visible activity declines and training events are not so prominent in the firm. It is assumed that management development has become an integral part of the organization with managers at all levels being committed to management development.

Career development

Each phase will have typical activities associated with the development process. For example, Phase II above proposes appraisal schemes and management audit, followed by training schemes (both internal and external). The other crucial activity here is that of career development which is closely linked with the appraisal process.

Careers can be developed so that the firm has sufficient site managers, senior surveyors and other necessary personnel to meet expected needs arising from growth or shift in workloads. This is, as we have seen, a reflection of the concerns of top management whereas individual managers will see career development in terms of making progress within the firm in terms of gaining interesting work, increasing responsibility and security. To the individual, such progress and increases are often associated with improved incomes which may be another important drive in career development.

Therefore, we have seen what activities in career development interact with management development. This interaction is illustrated in the model by Ashton and Easterby-Smith (1979). Reference to this model (Figure 7.4) shows that management development activities are to a greater or lesser extent determined by the information available from audit, assessment of managers in appraisal and the career development decisions of the manager and the organization.

Types of management development

Langford and Newcombe (1992) have classified management development activities into three types:

1. Formal management development.
2. Informal management development.
3. Integrated management development.

The formal management development process

Mumford (1986) sees formal management development taking place when there are changes in the job and in the job context. These changes are coupled with a programme of development activities.

Changes in the job context include promotion, job rotation, secondment and allocation of special duties such as committees or task forces.

The associated development activities which can help the individual manager to cope with and benefit from the change are:

Coaching – which can take the form of direct discussions and guided work to enable the manager to do a better job. It may also include 'shadowing' a more experienced occupant of a similar managerial position for a time or the 'sitting by Nellie' approach with one or more experienced people.

Counselling – may simply be the giving of advice about a particular aspect of the job or in a more complex form, dealing with any job-related or personal concerns, difficulties and anxieties which the job change has brought about.

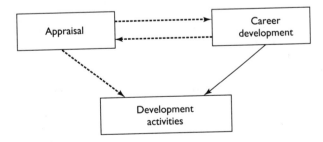

Fig 7.4 Interaction of management development activities

Appointing a mentor – a mentor acts as a parental figure to young, aspiring managers and oversees the overall development of the younger person's skills and job attitudes. However, as we have mentioned before, becoming a mentor can be seen as an aspect of management development for older managers who wish to pass on their skills and experience to others.

'Mentoring' has become a fashionable way of providing for development and has been shown to be beneficial to hitherto disadvantaged and discriminated-against members of organizations, women, ethnic minorities and the disabled. In particular, if the mentor is from the same group, a powerful role model is provided to encourage and inspire the individual who is seeking to overcome difficulties.

A mentor may also act as a coach and/or as a counsellor and as well as providing inspiration and a role model is also developing attitudes and behaviour associated with the whole of the career and the whole of the job. In other words, it is a general dispersal of a philosophy of management rather than coaching for better performance in one aspect of the job.

Coaching, counselling and mentoring are all part of in-house management development and they need to be supported by the provision of training courses. These can be either internal company courses or external courses. There is some evidence that the content and attendance on these two types of courses are dependent on the level in the organization: junior managers attend the in-house courses whereas senior managers attend the external courses.

In-company courses are directed towards specific company objectives and seek to integrate technical and managerial knowledge which is modelled on the practice of managers at work. Represented in some, but not all in-house courses are behavioural topics such as team-building, leadership and the management of change.

There are a number of reasons why the senior management of construction firms tend to rely on external courses; some of the reasons relate to issues of status and breadth of vision. Those who are status-conscious may resist attending courses which are taught or organized by those lower down in the hierarchy. Also, senior managers are more likely to respect outside experts than in-company specialists.

Breadth of vision and experience is a more acceptable reason for external courses in that, unlike internal courses, they provide a peer group for senior management. Such a group can together synthesize a wider view of the firm and take into account the social, political and economic environment in which the firm is embedded.

This wider view comes from the creation of 'networking' opportunities which should, ideally, provide a forum for managers to meet those in similar positions in other organizations. A word of warning, however: in a non-ideal way, such networks can, correctly or incorrectly, be seen by outsiders as powerful cliques and cabals bent on sinister purposes, secrecy and the exclusion of those who, for acceptable or unacceptable reasons, do not match the criteria for admission.

The informal management development process

Mumford (1987) says that informal development can be said to take place when 'the learning process is essentially subsidizing the activity of carrying out the job'. In the construction industry, where crisis management is more often the rule rather than the exception, such actual management development as takes place is seldom planned in advance. The intention of the organization, as well as getting the job done, is to enhance the individual manager's task performance.

An example of informal development occurs if a site quantity surveyor is charged with running a project because the firm has no other site managers to assign to the work. This change in job context enables the surveyor to experience a form of management development even though it may be of the 'being thrown in at the deep end' variety. Because of the suddenness and lack of preparation, this experience is likely to be unstructured and the development gains are likely to occur after the surveyor has run the job for a while.

The integrated management development process

This, as may be expected, synthesizes the informal (and incidental) process with the formal (and structured) process. This integrated model states two equal objectives:

1. There is concern for learning, and this develops the individual.
2. The task is done which satisfies the needs of the firm.

If the informal, incidental approach means being thrown in at the deep end, then the integrated approach means that the would-be 'swimmer' is also thrown a lifebelt and given thorough swimming instruction.

Another crucial factor is that the integrated process is planned with clear objectives. This means that the results can be evaluated and the management development process can be continued, enhanced or amended. However, the ultimate aim of the integrated process, and its most powerful advantage if it works, is that the superior and the subordinate negotiate the content and structure of the development process.

How management development is organized and costed?

Construction organizations are not typical of firms who have a strong commitment to management development although in the 1980s many of the large construction organizations began programmes of investment in management development, usually under the aegis of training departments. The size of the organization is crucial here. Mangham and Silver (1986) found that size was one of the important variables in the propensity of a firm to undertake management development. In other words, large organizations undertook more training and development activities. Important variables are:

- The level of technology within the organization. Those industries with high technology were more likely to undertake development. This begs the question of what is a high technology industry – one (debatable) criterion is that if the R & D budget is greater than one per cent of sales, then a firm is classified as a high technology company.
- The status and nature of the company. If it is owner-managed and a single company (no subsidiaries and not part of a group of companies), then it is likely to develop its management staff.

What seems to matter here is the willingness of a company to contemplate investing time and money in management development. Once they are willing to take the risk, they are agreeably surprised by the results. Mphake (1989) found that in 38 of the 100 largest UK construction companies, 17 had formal management development policies and these 17 also explicitly recognized the importance of management development to their company's efficiency and profitability. The question remains, however: what of the remaining 83 top companies?

The answer, unfortunately, is that many construction companies tend to see expenditure on management development as an unnecessary cost, which should be avoided, rather than as an investment. One way around this is to meet cost argument with cost argument as Kenney and Reid (1986) do. They regard investment in management development as any other form of investment. In other words, it should only be undertaken when a cost and benefit appraisal has been carried out. They suggest two questions which may be used to test the efficiency of management development:

1. What benefits can the organization obtain from investing in management development?
2. What costs are involved in the proposed development programme?

Benefits of management development

(a) The development of managers helps them to learn their jobs quickly and effectively, thus minimizing learning costs, and lessening the likelihood of expensive mistakes and the cost of absence due to stress.
(b) The present and future work requirements are likely to be met and maintained if there is management development in the organization. There is a responsibility on the part of the firm to be clear about its future work requirements.
(c) Following from (b), management development can become an agent for growth in an organization, helping both to stimulate and control expansion. Where growth has already started, management development should be an urgent priority if growth is to be sustained.
(d) The general morale of employees is improved by management development activities. Also, labour turnover among new managers, if caused by inefficient learning and inadequate development, can be reduced by staff induction, training and the coaching, counselling, mentoring activities.
(e) Connected with (a) above is the point that, with management development, managers can learn their jobs rapidly and are thus more likely to achieve higher levels of job satisfaction. This, in turn, increases the probability that they will remain longer with the organization.
(f) Linked with (c) above is the importance of management development in enhancing the ability of the organization to accept, cope with and implement change in a proactive rather than a reactive way by taking advantage of new opportunities. In order to do this, it must have staff whose knowledge and skills are up-to-date and contribute to the objectives of the organization. Management development enables new skills and knowledge to replace obsolescent ones.

However likely these benefits may appear to be, there are difficulties in valuing them which may explain the reluctance of companies to embark upon development programmes. Pepper (1984) says that a part of the difficulty is that the development process is fraught with uncertainty about whether development of talent really does take place and, if so, over what period?

In conclusion, the difficulty of quantifying benefits to a cost conscious industry is that some of the benefits, although vital are intangible and some may be more important and relevant to the individual rather than to the organization.

Costs of management development

The costs of development are more easily quantified, but there are different ways of looking at development expenditure. Listed below are three possible approaches:

1. To integrate opportunity costs into the costings. For example if a development activity is costed at £1,000 and then during this development the firm loses an opportunity to win a contract which would have rendered £10,000 profit, then the opportunity cost of the development is £11,000.
2. To minimize learning costs which are incurred by all construction organizations. Learning costs are payments which are made by organizations to staff who are less effective and efficient.

Examples of learning costs are:

- the costs of wasted materials or wrong decisions whilst learners are acquiring competence
- the reduced output of colleagues who have to partake in the training
- induction costs incurred when learners leave because they cannot meet the demands of the job

Although development costs reduce learning costs, all firms aim to minimize both, but, as with the benefits of development, learning costs are difficult to quantify which means that minimizing them is conceptual rather than practical.

3. To catalogue the three principal costs:

    - establishment costs
    - marginal expenditure costs
    - interference costs

Establishment costs are those which pay the salaries and operating costs (e.g. rental of office space, etc.) of the staff of the management group.

Marginal expenditure costs are those which add to the costs of providing a development function; for example, course fees, consultants' fees and individual expenses to attend events and courses.

Interference costs are likely to be the greatest because they are those which occur when a manager is not in his usual work station. Productivity may drop member of the management team is missing and a substitute manager may not be as effective or efficient, at least to begin with, although the experience may provide management development for the substitute manager.

Provision for management development

Now let us consider how the construction industry undertakes management development. As we have seen, only a small number of organizations do undertake it, but those that do place a lot of emphasis on it. We considered earlier the connection between management development and technical/professional qualifications and membership. It does appear that organizations with development programmes prefer to offer internal courses which are accepted as a qualification towards professional examinations. This may be because the construction industry regards technical knowledge as important and therefore expenditure on sponsorship for technical qualifications is seen as acceptable.

Although there is little evidence available, what there is shows that the use of technical courses is supported by coaching but that development techniques like project attachments, job rotation and self-directed learning are not widely used. Also, although internal and external courses tend to emphasize technical matters, there is an increasing emphasis on management techniques.

However, despite these encouraging trends, the budgetary allocation for management development is small. In 1985 the Industrial Society found that construction organizations spent less then 0.5 per cent of their annual turnover on development activities. Three years later, it was estimated that approximately £300 per manager per year is a typical figure for the provision of management development in the construction industry.

Table 7.3 Types of courses offered to construction managers

Courses	Percentage of courses for:		
	Junior Managers	Middle Managers	Senior Managers
Functional management e.g. finance, marketing, personnel, safety and training	33	35	33
Management techniques e.g. team building, IR, communications, delegation, motivating, leadership	58	51	42
General management e.g. business policy, strategy, decision making	9	14	25

(*Source:* Mphake, J. Management Development in Construction, in *Management Education and Development*)

136

It would seem though, that there is value for money in this expenditure, since the evidence concerning the methods used indicates a high level of activity (see Table 7.3).

The conventional focus of management development is the provision of appropriate courses. However, a large part of construction management involves communication and dealing with many and diverse people. Human skills and personal qualities are also important and have to be developed alongside the knowledge that a construction manager needs. There is therefore an imbalance in the skills being taught in management development courses and those of empathy and understanding are not well provided for.

Prejudices and barriers

If a company is to have a successful management development programme, it has to provide more than just a set of training courses, it has to make it a way of life and an integral part of the culture of the organization. This will then be reflected in the attitudes of all managers from the top to the bottom who will see the company as a place for learning and growing and not just an employer. However, this approach to management development does not arise by accident.

Development has to be deliberately designed into the strategy and fabric of the organization. This will not be easy, and there are some further considerations to be made.

Firstly, an obstacle that still has to be overcome is the old, but persistent (because comfortable), belief that managers are 'born not made' and another paradoxically connected belief that a person can be trained once and for all at the beginning of their career.

Secondly, it is imperative to question how much of management development thinking is predicated on the traditional masculine career path, i.e. period of education occurring at regular intervals during continuous employment until age 60–65, in the course of which the individual works his way up the hierarchy as far as he can go through positions of increasing responsibility and seniority. During this continuous, upward progress, the male manager is presumed to be solely occupied with his work role; home, family and leisure are marginalized or seen as irrelevant or distracting nuisances. This argument is amplified in the chapter on Women in Construction.

This is no longer acceptable for a number of reasons. One of these is that over the last decade research into occupational stress has shown that satisfying activities outside work, whether family, social or leisure oriented, are a necessary counterbalance to the pressures and challenges of management work. Those who do not have or do not choose to take up these necessary counter balances are in grave danger of succumbing to the physical and psychological ravages of occupational stress. Apart from the human cost, which can be life itself, the cost to organizations of sickness, absence and mistaken decisions caused by stress is very high.

Another important reason is one that affects the industry as a whole. If it is seriously committed to equality of opportunity for women – to employ, promote and develop women engineers, surveyors and managers – then it will have to think very hard about the flexible and interrupted career paths which, in the absence of and short-term unlikelihood of more radical social change, women have to follow. This means considering extended leave for the bearing and rearing of children, periods of part-time work and part-time study and refresher courses. If some of the more progressive companies could do this and maintain a high profile while doing so, then the industry would lose its negative image as a backward-looking male bastion and would be able to attract able and talented young women.

We have looked at the needs of the industry for managers with human empathic and understanding skills. Whether by nature, upbringing or both, these qualities are usually more highly developed in women, just as risk-taking and assertive behaviour are usually more highly developed in men. This not to imply a polarity between the genders; these are human not sex-specific qualities but those not considered appropriate in one or the other are suppressed so that men are inhibited in the expression of empathy and understanding and women in risk-taking assertion. A more equal balance of men and women could eventually lead to a more open expression of all human qualities. However, the construction industry still has a lot of work to do to convince women that it has more to offer them than menial servicing tasks and being the objects of scaffolding ribaldry and obscenity.

Thirdly, in discussing the role of the mentor, the matter of age becomes an issue. The decline in the relative numbers of young people in the 1980s and 1990s should make firms realize that they can no longer target their recruitment exclusively on the 16–35 age group. Older managers with experience in other industries or the services have a great deal to offer in terms of human skills even if they do not have the technical knowledge of the industry. Such people could act as mentors or as management development consultants in a full or part-time capacity.

Part-time work, as well as being essential for women, could also be very beneficial to older managers and greater flexibility in hours of work and retirement ages could benefit both managers and companies. The distressingly high morbidity amongst managers who move from full-time work to full-time retirement could be lessened by a gradual reduction in hours worked to enable adaptation to and preparation for retirement and, if he so wished, a manager could be retained beyond retiring age as a consultant, advisor or mentor.

Fourthly, the models of management development are very much rooted in British and American management theory. Increasing involvement in Europe means a wider perspective must now prevail and, whilst cultural differences must be taken into account, each country should be able to learn from the others.

In other parts of the world, self-determination rather than the imposition of alien and inappropriate methods is important. The development of construction managers for Africa and Asia should take what is appropriate from the Anglo-American models, discard what is not and build in its place the activities and skills which are necessary to and compatible with the culture in question.

Finally, a still emerging challenge is that of Eastern Europe. The more progressive elements in these countries are looking to the West for management thinking to up-date industry and provide construction services.

This may be the management development challenge of the 1990s.

Questions

1. Identify and evaluate the most important components of the management develop process for the construction industry.
2. What are the main obstacles to management development in the construction industry?
3. Develop a methodology appropriate for the construction industry by which the costs and benefits of management development may be evaluated.

References

Ashton, D., and Easterby-Smith, M. (1979) *Management Development in the Organization: Analysis and Action*, Macmillan

Ashton, D., Easterby-Smith, M. and Irvine, C. (1975) *Management Development: Theory and Practice*, Bradford Management Bibliographies and Reviews

British Institute of Management (1969) *Management Development and Training: A brief survey of the forms used by 278 companies*, BIM

Buckley, J. and Kemp, N. (1987) The strategic role of management development in *Journal of Management Education and Development*, **18(3)**

Construction Industry Training Board, *Annual Report, 1989*, HMSO

Fryer, B. G. (1979) Management development in the construction industry in *Building Technology and Management*, 11 May, pp. 16–18

HMSO, *Housing and Construction Statistics, 1989*, HMSO

The Industrial Society, (1985) *Survey of Training Costs*, The Industrial Society, London

Kenney, J. H. and Reid, M. (1986) *Training Interventions*, Institute of Personnel Management

Leggat, T. W. (1972) *The Training of British Managers: A study of need and demand*, HMSO

Langford, D. and Newcombe, R., (1991) *Competitive Advantage in Construction* (eds. Male, S. Torrance, V. and Stocks, R.) Butterworths

Langford, D. and Newcombe, R. (1992) Management Development in Construction, in *Competitive Advantage in Construction* (eds Stocks, R. and Male, S.) Butterworth

Manpower Services Commission (1981) *Glossary of Training Terms*, HMSO

Morris, J. (1971) Management development and development management, in *Personnel Review*, **1(1)** pp. 30–43

Mphake, J. (1989) Management Development in Construction, in *Management Education and Development*, **18(3)**, pp. 223–243

Mumford, A. (1986) *Handbook of Management Development*, Gower

Williams, R. (1987) Management training for civil engineers in mid-career, unpublished MSc thesis, Brunel University

Chapter 8

People and information

Simon (1960) believed that management equated to decision-making whilst many other management authors, the people school, believed that people are organizations and, hence, both the active constituent and the focus of management. In this 'information age' it is a short step to suggest that management is making decisions about people. People are clients, customers, owners, capitalists, managers, workers, etc; as the roles of people are multi-faceted, so are the inputs to and consequences of decisions.

Situations are complex and it may be that only change is certain. The development of information technologies has had a major impact on decision-making – hence on people's abilities to take factors into account in making decisions and forecasting consequences; people are required to operate the machines. Hence, people are affected in a wide variety of ways.

The decision support model

So, if the focus is on making decisions, it is important to consider the decision-making process and the decision-making environment. A simple systems model illustrates the process and the environment of decisions.

The environment may be considered as a multitude of layers (rather like an onion). As it is important to consider the boundary condition – such as for deciding how open (or closed) the system is or how readily the system responds to changes in the (external) environment – it is essential to identify where the boundary lies. The boundary location will be determined by the decision level – project, organization, industry, etc. This consideration is important in identifying the realm of the decision – what is controllable and what is not; dependent and independent variables, endogenous and exogenous factors.

A common purpose in identifying the boundary is to determine PEST factors in the environment (Political and legal, Economic, Social and Technical) to input to SWOT analysis (Strengths and Weaknesses – internal; Opportunities and Threats – external).

Data and information

Data, facts and figures, usually must be processed into information which is comprehensible and, hence, useful to people in making decisions. The determination of what

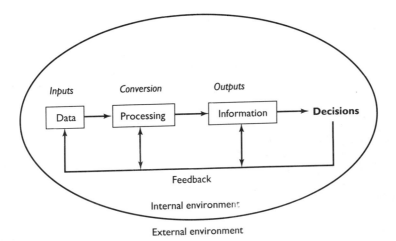

Fig 8.1 A systems-based decision support model

information is necessary is governed by the decision, so the data to be obtained, and the processing required to produce information from it, are driven by the decision. Thus, the objective of providing information is to support decisions.

Occasionally IT machines make decisions but they do so based on rules given to them by people. The more usual situation is for the machines to provide information to assist people in making decisions. People remain the active, thinking constituent.

Fear of information systems

Unfortunately, it is common for people to feel threatened by the introduction of information technology or the upgrading of existing systems. Generally, the fears fall into two categories:

1. Fear of inadequacy.
2. Fear of unemployment.

As fear of unemployment is more precise, it may be dealt with more easily. Overall, IT has been an employment generator but it does involve the abolition of many mundane, data storage jobs (filing) – the sort of tasks which computers do very efficiently and effectively. In fact, the taking-over of such tasks by computers often has meant that the people 'released' have been employed on other, more fulfilling jobs; however, some undoubtedly have become unemployed as a direct result of IT, despite the creation of other job situations.

Fears of inadequacy tend to be both quite complex and hidden/disguised. Usually, people are reluctant to admit and reveal differences (whether actual or perceived). It is in this sphere that much reassurance may be necessary and the need for education/training is great.

Frequently, information management and information systems are regarded as being synonymous with information technology. Although such congruence occurs

141

increasingly often in practice, it is far from a requirement; many hand-based information systems operate very successfully and, perhaps especially for small construction organizations, are more suitable (in terms of efficiency, effectiveness and human impacts) than their IT counterparts. However, it must be noted that the permanence of such a situation is open to increasing question as the technology advances on an increasing variety of fronts!

Raw data

As a first approximation, it is reasonable to consider data to be raw facts and figures. The data may be internal to an organization or project (it has been generated by and collected/recorded by the organization – such as measures for production bonus payments) or external (such as data published by the Central Statistical Office (CSO) in housing and construction statistics or the Building Cost Information Service (BCIS)). However, 'raw facts and figures' implies accuracy and it is vital to appreciate that data are not exact – they are subject to errors and, potentially, bias. Errors are random mistakes; bias is systematic distortion, either intentional or unintentional.

It is common for a more detailed database to be regarded as more useful, flexible, etc. However, although that may be true for a variety of purposes, it may be negated by a higher level of errors in allocating data (measures) to the items (categories) constituting the database as well as increased error in collecting the data.

Fine (1975) reported a study carried out at the Building Research Establishment (BRE) in which site accountants were asked to allocate site costs to databases of varying details. The results of the allocation exercise were:

- 30 cost categories; 2% of data misallocated
- 200 cost categories; 50% of data misallocated
- 2000 cost categories; 98% of data misallocated

Sometimes misallocations may be of trivial consequence but such misallocations may produce serious errors due to error accumulation!

The likely trade-off between detail and accuracy of a database is clear so it is important that both these factors be assessed in designing the database. Possible bias must be considered too. Likelihood of bias can be reduced by ensuring independence of data collectors, processors and users as well as using simple collection techniques employing pro-formas or 'automatic' means. Clear, concise descriptions of each item in the database are essential.

Data are not free. Some may be purchased, some may be obtained by commissioning an agency to produce them for a fee and others may be obtained internally by paying wages/salaries and other costs of employment, overheads, etc. for their provision by employees. Irrespective of the source, time is involved too. Obtaining data will be subject to diminishing marginal costs (while their use may be subject to diminishing marginal returns). McDonagh (1963) estimated that information cost as a proportion of the costs of running an advanced economy was over 50 per cent whilst Koontz et al. (1984) put the proportion at around 20 per cent. As much of the cost will be due to data collection, it is advisable to assemble data so that they may serve several purposes (i.e. multiple use databases).

Data must be adequate; this depends upon the amount, variety and 'depth' (quality) of the data. Measurement of adequacy is judgemental and should consider the multiple uses of the data (transferability) as well as suitability to supply information to support individual decisions.

Hence, overall, it is appropriate to evaluate the sufficiency of data – combination of the adequacy and accuracy of the data within the constraints of its provision and use, notably time and cost factors.

The information which is output by the system is either historic or predictive. Historic information is a record of past events and may include interpretation of them. Predictive information are forecasts and so incorporate the problems inherent in the data on which the forecasts are founded and those in the forecasting technique(s) employed.

The most common forecasting technique used in the construction industry is to extrapolate trends; modifications of the predictions are made to incorporate likely future events from publications (e.g. government policy statements) and the experience of the forecaster. Whatever technique is adopted, forecasts should aim to be:

- accurate – minimal errors
- unbiased – equally likely to over-predict and under-predict
- efficient – have the smallest variance for the sample size
- consistent – errors reduce as sample size increases

Forecasts are used to aid decision-making by predicting what is likely to occur in prescribed circumstances. Hence, forecasts are used to help decide what to do but also provide a basis for control by comparison of the actual outcomes with what was predicted. Unfortunately, two major assumptions often are made – that actual performance has been measured accurately and that the performance predicted is correct (notably, that the circumstances have not altered). For proper control, it is important to take account of inherent variabilities so that action is instigated only when necessary, or when correction is required as performance lies outside the margins of error (variability). Many information systems, therefore, are designed to produce exception reports – to report performance only when actual differs from predicted by a predetermined amount. Such reports highlight areas for managerial scrutiny and action.

Data variability

Information varies in nature from very formal (such as a company's annual report and accounts) to highly informal (such as a casual conversation over lunch). The more formal and permanent (e.g. written and published) the information, the more authoritative it appears. However, informal information may be extremely useful in amplifying and explaining formal items. Also, formal information may take a considerable time to produce and so informal information often is available earlier and, as such, is more useful.

The importance of informal information and communications in the construction industry was highlighted by the Tavistock Institute in 1966 and 1967. However, care must be taken over rumours and similar informal information – it is important to verify such information before acting on it, especially if the actions are quite drastic (such as actions on rumours of impending bankruptcy - these may become self-fulfilling prophesies!).

Data durability and visibility

Information varies in durability from the viewpoints of the time over which the information is valid and, often more important, the time for which the information is useful to support decisions. For managerial control, reliable information must be received as early as possible to maximize the possible control; although information about performance of a completed operation may be helpful for planning the operation when it occurs again, that information is useless for controlling the operation on the particular occasion.

As information provides visibility, it gives power to those who possess it and so must be controlled. No one has complete information, so people must make decisions and act within the boundaries of the information available along with other constraints. As people are believed to behave rationally in trying to achieve their objectives, a common description of decision-making behaviour is 'bounded rationality'. By providing visibility, information aids delegation and identification of individuals' responsibilities; commonly, access to information mirrors the structure (hierarchy) of an organization. Thus, restricting access to information establishes and maintains power of those who have the information. Informal contacts, relationships, communication channels, etc. are used to circumvent restrictions.

Data quantity and quality

Naturally, problems arise from lack of information but too much information (overload) and inaccurate/distorted information may be at least as problematic. Overload may be combated by providing decision-takers with appropriate information only (in quantity and scope). Inaccuracies and distortions, especially if deliberate, may be very difficult to detect, particularly early detection. So, potentially, errors are extremely dangerous.

An important safeguard, notable where computer-based systems are used, is to ensure that those who use the information have sufficient knowledge/expertise to detect likely errors – not in detail but to know that the information (such as the solution to an engineering problem) is in the right 'ball park'. The approach helps to overcome some 'computer blindness' difficulties too for, although the computer may remain a 'black box' to the user, there is usually a built-in checking mechanism which invites user interactions.

Communications

If information is power, communications allow the power to be used! Communications is about interactions – the transfer or display of data and information, ideas, knowledge, etc; it is the process of transmitting a message by using symbols from sender(s) to receiver(s). Efficient and effective communications are vital. Output from one person is input for another so it is necessary to consider both the transmitter and the receiver as well as the medium between them.

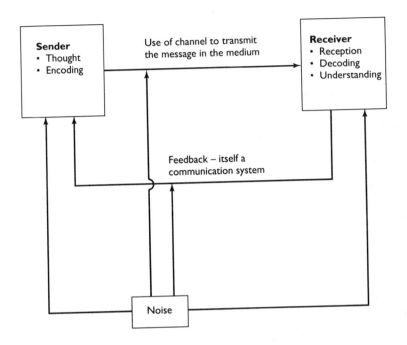

Fig 8.2 A communication system

Construction is a labour-intensive industry, each project having inputs from a wide variety of disciplines and organizations. Technology plays an increasing role in information provision, so good communications are both vital and at the same time complex. Commonly, communications occur 'by tradition' ('I've always done it that way') or by following the contractual or organizational dictates. Despite advantages due to standardization and familiarity, traditional and contractural methods of communications can have the problem of conservation – perpetuating the *status quo* and discouraging (if not excluding) new/alternative approaches.

Whether a communication appears to be between computers, organizations, individuals or some combination, it is the people interface and actions which are essential (in programming a computer, writing a letter, giving a verbal instruction, etc). Therefore the principles of good communications are universal. Figure 8.2 shows a communication system. The sender's desire(s) must be expressed in a language comprehensible to both sender and receiver and in a form suitable for transmission. Hence, the sender must consider the abilities of the receiver and match those to his/her own as well as selecting the most appropriate medium and channel for transmission of the message.

Problems in communications
No communication will be perfect due to interference – distortions within the operation of the system and 'noise' from the system's environment.

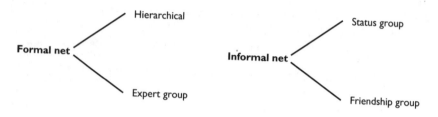

Fig 8.3 Formal and informal communication nets

Feedback is important in order to check that the intended message has been transmitted correctly; but the feedback loop is a communication system itself. Especially for long, complex messages, the effectiveness of communication can be enhanced by using several channels/media, for all or parts of the message, to obtain reinforcement. Such an approach will assist in highlighting areas of ambiguity, errors and omissions. Of course, all the efforts of the sender will be to no avail if the intended receiver is unprepared or unwilling to receive the message (data cannot be entered into a computer which is not switched on!).

Problems in communications arise in a variety of ways and from many causes – interference, omissions, ambiguities, languages (codes), etc. are quite obvious. Often, such problems can be detected easily and measures taken to remedy the situation. Most construction contracts acknowledge the likelihood of such problems occurring occasionally and make appropriate (procedural) provisions for their solution (e.g. JCT 80 clauses concerning omissions of information, late provision of information, discrepancies/divergencies between contents of contract documents). However, other areas of possible problems are less obvious – some, by their very nature, may be 'designed' to be difficult to detect – distortions/omissions by the sender (accidental or deliberate); perceptual bias by the receiver. Overloading people with messages makes it very difficult for them to discover important items. Distrust between people often results in censorship/screening of messages.

Especially where there is conflict between messages, it is useful for there to be a message hierarchy. Commonly, the hierarchy is derived from the status of the people involved, notably the sender, from expressed hierarchy (as in JCT contract documents) or from the form of the message – written messages taking precedence over oral ones etc.

Overcoming problems in communications

Handy (1985), provides three principles for encouraging effective and efficient communications and avoidance of communications problems:

1. Use more than one communication net (network).
2. Foster two-way communications.
3. Keep communication chains as short as possible.

Formal communications usually follow an organization chart or accepted system (such as in formal meetings) whilst informal communications are shaped by relationships between individuals, as demonstrated in Figure 8.3.

Figure 8.4 shows some common communication networks. The chain and star (or wheel) are hierarchical in which the 'leader' collects, collates, judges and uses the information. People at the periphery are likely to feel isolated from decisions and highly subservient; others may perceive themselves to be mere 'postboxes' without any real input to or influence on decisions. The all-channel and circle are egalitarian networks in which everyone can be involved.

Communication networks

Although much work on communications has concerned small groups in controlled situations, it appears that hierarchical networks are most suited to routine and

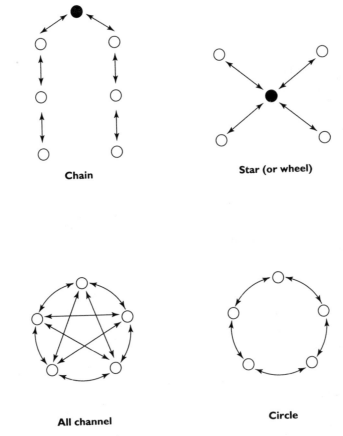

Chain

Star (or wheel)

All channel

Circle

Fig 8.4 Communication networks

bounded (defined) problems whilst egalitarian networks, due to their potential to generate novel, creative methods and solutions (e.g. 'brainstorming'), are appropriate for unusual and unbounded problems. Egalitarian networks tend to operate more slowly and erratically so may be more appropriate for design situations whilst hierarchical networks may be more suited to the construction phase of a project.

Authoritarian organizations emphasize vertical communications, especially downwards. Due to lengths of chains, it is easy for messages to be distorted or, even, lost. Good feedback is vital to ensure the correct message reaches the intended recipient (and necessary action followed). Katz and Kahn (1977) noted five main types of messages communicated down hierarchies:

- directives for executing tasks
- information to help understanding of relationships between tasks
- information about practices and procedures
- feedback about subordinates' performance
- information about the organization's goals

Upward communications may be hampered by filtering to suppress unfavourable messages and telling superiors what it is believed they wish to hear.

Good vertical communications are important to keep people in touch; to retain commitment by fostering belonging – suggestion schemes, 'open door' policies are helpful.

Crosswise (horizontal and diagonal) communications will occur in a hierarchical system, both formally and informally, between people who are not in direct reporting relationships with each other. Such patterns speed the flow of information and general awareness but may involve problems of flows of inappropriate information (hard-to-keep secrets), people exceeding their authority and going direct to the 'doer' so that a superior is not kept informed.

A key factor in good communications is to be concise; supply the required detail only in a way appropriate for the recipient. Diagrams, charts and graphs are very useful in providing succinct appreciation of situations, trends, etc; they can be backed up with tables of data, if necessary, and with text to provide discussion.

Often, the first real sign of things going wrong on a project, or in an organization, is a breakdown in communications; perhaps manifested by a rapid abandoning of informal channels for formal ones and increasing 'officialdom' (such as quoting contract clauses).

Information systems

Information systems are the means by which information is provided to support decisions. Hence the system is required to collect and produce data and information in order to facilitate efficient and effective operation of the organization, ensuring compliance with statutory (and similar) requirements. Thus, to provide visibility!

In choosing between alternative systems, a variety of questions should be addressed, including:

- What is the purpose of the information?
- When is the information required?
- Who should receive and use the information?
- Why is the information necessary?
- Where is the information required?
- How is the information wanted?

Attention must be given to feedback, the accuracy required, the levels of detail and communication mechanisms.

The ability to respond to change

An important aspect of the system is its degree of openness (glasnost!?) – how responsive is the system to changes in its environment? This aspect involves the PEST factors and SWOT analysis. At the open extreme of the spectrum of possible boundaries for a system, the system responds immediately and 'totally' to changes in its environment whilst at the 'closed' extreme the system's operation is isolated from its environment. Most information systems are quite open – in most cases that is a major requirement/purpose of the system (although the technology of the system (computers etc) must be in a protected, isolated environment to operate properly).

Hard-wired communication networks

Many individual computerized information systems are physically linked to others to produce a network. These include:

LAN	Local Area Network	Access points/facilities on a single site only
WAN	Wide Area Network	Access on several sites
VAN	Value Added Network	Incorporates storage, processing etc as well as communication
VAS	Value Added Service	Communication and information services supported by VAN
VANS	Value Added Network Service	Combined VAN and VAS

Reactive or proactive data information systems

For efficiency, waste must be eliminated so all information provided must be necessary to support decisions and corrective measures. In providing information, systems operate in either reactive or proactive modes. The reactive mode is where the system responds to requests/requirements for information whereas the proactive mode involves providing information in advance of management expressing a request. Much information is provided regularly/continuously (e.g. weekly costs and outputs) as part of the routine management within an organization.

It is important that an information system has sufficient capacity to accommodate an appropriate balance of operations – notably that urgent requests for information (reactive mode) can be dealt with promptly without detrimental effects on other aspects of information provision.

Management control through data information

The information cycle, Figure 8.5, shows how the information system can facilitate managerial control; it should do so at a variety of levels. Operational control involves everyday operations (e.g. payroll). Transaction processing systems (TPS) are used which do not, themselves, yield information for other levels of managerial action but may be integrated into other systems which do.

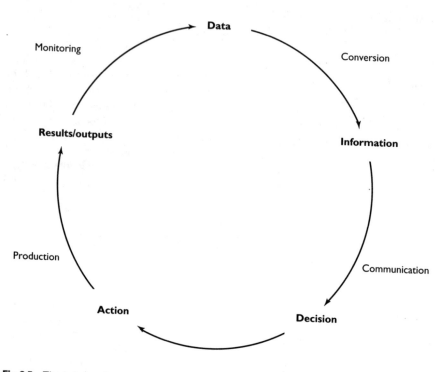

Fig 8.5 The information cycle

Management control involves short-term decisions; Information Provision Systems (IPS) use outputs of TPS to produce reports, etc. A common aid to management control is the production of trends by the system.

Strategic control is long term. Decision support systems (DSS) use global models of 'the world' and, in particular, help to answer 'What if?' questions. Often DSS use heuristics to produce forecasts of consequences of alternative decisions.

Programmed decision systems (PDS) make decisions automatically in accordance with heuristics specified by management. Of necessity, such decisions are straightforward.

Expert systems or intelligent knowledge-based systems (IKBS) try to emulate human decision-making by following prescribed decision rules. Systems which 'learn' from past decisions and from additional data are being developed.

System selection

The provision of a management information system is driven by the needs of the users, that is the outputs as governed by the decisions which must be supported. Outputs of a system may be placed in the categories of:

- mandatory – to comply with the law
- essential – to enable the organization to function
- desirable – to assist/enhance the functioning of the organization
- supplementary – likely to be helpful

Especially during recessions, a great deal of emphasis is put on assessing constraints – particularly financial. However, for information systems, constraints involve hardware, software, legislation, etc. and, probably most important, the attitudes and abilities of the people involved.

The overall process of selecting an information system is analogous to that for procuring a building (or any other major project). Hence, the process begins with an assessment of requirements followed by a feasibility study to determine whether the requirements can be met within the constraints imposed.

Feasibility study

Outline sensitivity analyses may be helpful at this juncture. It is useful for the feasibility study to consider the horizon of developments as it may be worthwhile tolerating delay in acquiring a system to secure developments which ensure the system meets requirements at a higher level, thereby avoiding early updating, etc. of the system.

The feasibility study is likely to involve dialogue between prospective system suppliers and the acquiring organization. Hence, it is important to ensure that 'salespersonship' does not become dominant – it is the system which must serve the organization, not vice versa! An audit of knowledge and skills required to operate the (alternative) system(s) is helpful so that the skills of the organization's personnel can be evaluated against system requirements and, hence, any training requirements established.

The outcome of the feasibility study should include a performance specification for the system against which selected suppliers may be invited to submit priced schedules of proposals, as noted below (from Hodge *et al* 1984), for evaluation and selection.

1. *Introduction and Contents*

2. *Instructions*
    - Objectives and scope:
        - objectives
        - current system
    - Contacts:
        - co-ordinator
        - method of inquiry
    - Timetables:
        - briefings
        - benchmark
        - proposal due date
        - contract award date
        - installation date

3. *System Requirements*
    - Mandatory and desirable:

- cost
- time
- volume
- growth
- general configuration: hardware, software, etc.
- training
- contract clauses
- MTTR, MTBF
- current contracts
- service

4. *Evaluation Methods*
   - Service
   - Personnel
   - Training
   - Cost
   - Contract
   - Hardware characteristics
   - Software characteristics
   - Financial position

Acquisition of bespoke systems may involve particular difficulties in that the systems are individual and so may not be compared easily. Adequate time must be allowed for evaluation; it is essential to 'get it right'.

Determining the system specification

The instruction to suppliers should convey what the organization seeks from the system. Good contacts are important – one individual should act on behalf of the organization to ensure co-ordination and assimilation of information.

Timetables are vital to secure the system, and its installation, as required. Setting key dates for installation is very useful when changing to the new system smoothly and in sensible stages whilst the old system may be retained for temporary back-up (essential during testing).

Specification of the requirements for the new system, categorized as above, is essential for evaluating alternative proposals – this enables a 'value-engineering', cost-benefit approach to be taken.

Any evaluation should include cost – capital, maintenance and operating – and so may be executed by using discounted cash flow techniques. Time concerns both the period necessary to procure the system and the system's speed of operation (i.e. time to carry out major, specified functions). Volume considerations include capacities of hardware and storage facilities – regard should be paid to peaks and troughs of operation as well as 'average' workload. Provisions for and ease of extending/upgrading the system should feature in the evaluation – many systems grow incrementally by 'modular' additions.

Reliability and maintainability

Maintenance is an important factor in selection of computer-based information systems. Two important measures are Mean Time to Repair (MTTR) and Mean Time

Between Failures (MTBF). MTBF provides insight into the frequency of failures – hence, the in-use reliability of the system whilst MTTR indicates (the combined effects of) the seriousness of failures and the efficiency of the maintenance organization in effecting the repairs – hence, a measure of system 'down-time' is provided. For assessment of these factors, note should be taken of what is deemed to constitute a failure and when a repair is completed; MTBF and MTTR are helpful relative measures for evaluation of alternatives but rather dubious absolute measures.

Attention should be given to the types of failures so that their likely consequences can be assessed – this will be influential on the decisions of what back-up to provide. BS 3811 categorizes maintenance into:

Planned preventative running maintenance – minor items of work which do not interfere with the system's operation (e.g. dusting a keyboard).
Planned preventative shut-down maintenance – minor work items but which require the component to be taken out of service during the work (e.g. replacing a printer's ribbon/cartridge).
Planned corrective breakdown maintenance – failure of a unit but, due to planning for the failure, repair can be effected quickly (e.g. replacing a 'blown' fuse).
Unplanned maintenance – usually corrective of quite major faults which have low probabilities of happening (e.g. replacement of a failed disc drive).

Most computer systems suppliers offer a maintenance service. A contract for maintenance service often costs in the region of 15–25 per cent per annum of the capital cost of the equipment (hardware) to be maintained. Hence, what is to be subject to a maintenance contract merits scrutiny as does what the organization can afford not to be: this involves evaluating consequences of failure.

The typical failure pattern is depicted in Figure 8.6.

Non-threatening approach to change and training needs

Probably the most important aspect of selecting an information system, and one which all too often receives inadequate attention, is to consider the human factors. What do the people in the organization want and need? What skills do they have and how do these match to the requirements of the systems under consideration? What training will be necessary, how will it be executed, by whom, when and at what cost? Such issues may be addressed by undertaking an audit of skills, etc. to match abilities to requirements as well as unobtrusively and (very important) as unthreateningly as possible. The result would be a type of variance analyses which would indicate training needs, etc.

People, as the paramount 'resource', must be involved in a way which is positive for them; this helps them to evaluate the changes for themselves and to identify with those changes. As almost any change is perceived to be somewhat threatening, great skill and care is required in managing change to an information system. Information is the oxygen of any organization – it can't survive long if starved of (good) information. The greater the degree of change, the more difficult the management of that change is likely to be, the most difficult obviously being from a manual information system to a computer-based one.

Normally, best practice requires the system supplier to participate in evaluation of the training needs early in the changeover and to provide a comprehensive training programme. A convincing approach of helping employees is essential – that must be the purpose of the training. Managers may be overlooked but they need to be made aware of how the new system operates and what its capabilities are so that they can use the system to its full capacity but not demand unfeasible tasks of the operators! Refresher courses and periodic updating should be part of the agreed training programme.

The system must be right for you

Factors affecting selection of a computer system are shown in Figure 8.7. Naturally, all evaluations must be objective and use appropriate techniques. Often, evaluations are carried out in a biased environment because it has been decided that a new information system is needed and it must be computer-based. Whilst computer-based systems are, increasingly, appropriate, they are not a universal panacea especially for small organizations. Computers are expensive show-pieces/'toys' – they do what they are told to do by people; it is the people who make them work efficiently.

System implementation

Physical installation of a new, computer-based information system is likely to require building work. Preferably such work should be determined and executed prior to delivery of the new hardware. The work will be in respect of security for the system,

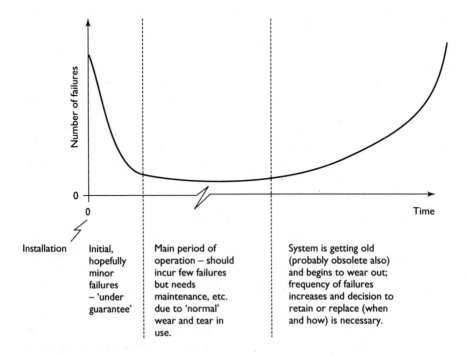

Fig 8.6 Typical failure pattern over life of system

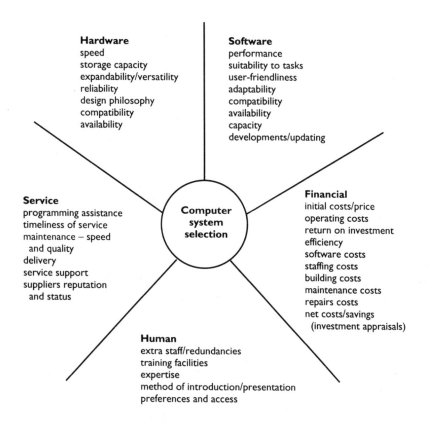

Hardware
speed
storage capacity
expandability/versatility
reliability
design philosophy
compatibility
availability

Software
performance
suitability to tasks
user-friendliness
adaptability
compatibility
availability
capacity
developments/updating

Service
programming assistance
timeliness of service
maintenance – speed
 and quality
delivery
service support
suppliers reputation
 and status

Computer system selection

Financial
initial costs/price
operating costs
return on investment
efficiency
software costs
staffing costs
building costs
maintenance costs
repairs costs
net costs/savings
 (investment appraisals)

Human
extra staff/redundancies
training facilities
expertise
method of introduction/presentation
preferences and access

Fig 8.7 Factors affecting computer system selection

environmental control, power supplies and communications wiring etc. Installation itself includes:

- delivery, set-up and testing of hardware
- installation, testing and de-bugging software, notably for bespoke software
- testing the whole system under operating conditions
- installing and testing security measures
- staff training/recruitment
- transfer of existing files
- documentation – control and user procedures

The testing of hardware and software should be done individually and then in combination under simulated operating conditions for the system. Indeed, rigorous testing should continue under very severe operating conditions. The testing should demonstrate how the system is likely to respond to heavy loading and potential problem areas and modes of failure. An objective should be to ensure that the system is unlikely to 'crash'. (Whenever a system crashes, invariably it does so at the most inopportune time!)

Methods of changing over to the new system

The change to using the new information system may be effected by cut-over, parallel or phasing-in methods. Cut-over involves total switching into the new system and abandonment of the old one once installation of the new system has been completed. Parallel involves operating the two systems side-by-side for a period; this facilitates comparisons and provides for building confidence in the new system whilst back-up is available from the old one. Phasing-in occurs on large systems where change to the new system can be executed incrementally.

Whichever method of changing is adopted, it is essential to identify the point at which the new system is handed over by the supplier for, at this point (as in construction projects), major responsibilities transfer from the supplier to the purchaser.

Whether the change is to a computer system or not, it is important to ensure that data from the existing system are transferred into the new system accurately and completely. Transfers between computer systems may be done automatically (or, at least, electronically) by use of a transfer package. Irrespective of the mechanism used, checking that the transfer has taken place correctly is vital and best done by examining the data transferred in small batches (as for original inputting).

User documentation

Good user documentation is essential to the efficient and effective operation of an information system. Hardware documents usually concern technical information. Of more concern to most users is software documentation – program manuals and introductory guides.

Manuals need to be adequate and detailed. Content needs to be fairly comprehensive but clear and simple so that they are 'usable' by the 'average' operator. A good contents list, index and updating service are vital.

System operating procedures should be provided and control documents should be used to monitor the system's performance. There should be a section concerning error messages with explanations, likely causes and what may be done to overcome the problem/correct the error. 'Help' sections are important to supplement the 'help' portion of programs.

As systems usually evolve with use, it is essential to keep documentation up-to-date. Commonly, an updating service for both software and documentation is available from suppliers but, for in-house programs, an individual should be given responsibility for the task.

Introductory programmes are very useful to help people learn how the new system operates, its capabilities (and 'quirks') and, perhaps especially, so that people can gain confidence in the system's operation and their abilities to use the new system.

Need for training of managers

It is important that managers who use outputs (information) produced by the system are aware of its operation as well as those who will be 'hands-on' operators.

Non-threatening approach to installations

Staff should be reassured throughout implementation, as almost all personnel are likely to feel at least a little threatened and to secure the commitment of staff to make the system work well. The commonly-perceived threats which need to be addressed may be classified as:

Organizational – 'big brother' notion.
Confidence – people tend to prefer existing methods, etc. New systems may imply lack of confidence by management.
Competence – new systems may require new skills.
Status – the status structure may be altered with the introduction of a new system (information is power!).
Health – many health problems have been cited as being caused by VDUs. Evidence of causation does seem to be tenuous but codes of practice are being developed concerning eyesight, posture, skin problems and use of VDUs by pregnant women.

The EEC has issued a directive on work with Display Screen Equipment following which the UK government has produced *Display Screen Regulations and Guidance* (DSRaG) which applies to all new equipment (effective 1 January 1993); existing equipment must comply by 1 January 1997. The objective is to protect employees who are habitual users of display screens (VDUs) and workstations (including keyboards, etc). The requirements include changes in activities of habitual users during each day, appropriate eyesight tests and provision of training and information. The Health and Safety Executive has also published information to provide advice over the use of equipment, such as *Visual Display Units* (1983).

Ongoing system improvement

Once in use, performance of the system should be monitored and evaluated against the specified performance. Shortfalls in performance should be rectified (if possible) by the suppliers. An additional facet of such monitoring and evaluation is to determine what upgrades are required and when. As 'work tends to fill the time available' so the maximum information tends to be sought and provided. Extensions/upgrades should meet only real needs – those which will benefit the working of the organization.

Responsibilities for monitoring and evaluations must be clear and the tasks must be carried out objectively. The work requires considerable knowledge of information systems, of the organization and its methods and philosophy of operating.

Safety and security

Safety and security involve considerations of hardware, software, data and information and people. Protection for people has been noted in the context of potential health hazards whilst most other facets of safety and security involve protection from people.

Apart from the 'normal' physical hazards of fire, etc. computer systems often must operate in a particular, controlled environment, isolated from major vibration, etc. Sunlight and magnetic fields are notable hazards for magnetic storage media (tapes, discs).

For security, attention must be given to physical access to the equipment as well as (remote) access to the system's contents. Both physical loss (theft) and damage and corruption of the system and its contents must be avoided. Hence, the problem which emerges is that of ensuring adequate access for uses whilst maintaining appropriate security!

Physical security can be achieved through appropriate siting of the equipment with lockable doors (card or numerical entry keys). Commonly, system security is obtained through logging-in procedures and use of passwords; by such mechanisms (protocols) different users obtain access to only the parts of the systems relevant to their jobs.

Systems content security is enhanced by backup stores which are not used for normal operations but retained separately in secure locations and brought into use in the event of a failure in the primary system.

Security of personal data

A further, and particular, facet of information safety and security is the protection of people and organizations, etc. about whom data are collected and stored. In the UK, following EEC initiatives, the Data Protection Act, 1984, took full effect on 11 November 1987. The Act applies to all automatic processing of data which relate to living people with the objective of enforcing standards of data processing and so providing protection for the individuals. The Act applies to all data users who control the contents and use of data from within the UK. It does NOT apply to data on manual files and systems.

The Act requires that personal data be:

- obtained and processed fairly and lawfully
- held for specified and lawful purposes
- neither used nor disclosed for purposes incompatible with those for which the data are specified as being held
- adequate, relevant and not excessive
- accurate and up-to-date
- kept only for as long as necessary

People are entitled to have reasonably prompt and inexpensive access to personal data held about them and, if necessary, to have the data corrected. Those who hold personal data must take adequate security measures to prevent loss of, or unauthorized access to, the data.

Data users and computer bureaux holding/using personal data must be registered for those purposes with the Data Protection Registrar. Some personal data (e.g. domestic, national security) are exempt and users of only exempt data are not required to register. Registration details include the name and address of the user, description of the personal data to be held, the purposes of holding the data, sources of data, disclosure and transfer details and the address to which people may apply for access to data held about them.

Policing registration and use of data is the responsibility of the Registrar. Generally, breach of the Act is a criminal offence.

Individuals have the right of access to data held about them and if an individual suffers due to:

- loss of data
- unauthorized destruction of data
- unauthorized disclosure of data
- inaccurate data

that individual may claim compensation from the data user. Reasonable care by the data user, however, is a good defence to such a claim.

Summary

This chapter considered the collection of data, their conversion into information, communication mechanisms and the use of information to support decisions. Although technologies are being used increasingly, the technologies work on the instructions of and under the control of people. Such 'master-servant' relationships should not be overlooked!

Decision support is the objective of information provision. The information must be adequate, appropriate, accurate, timely and not overloading. More sophisticated, detailed databases may be partly self-defeating due to problems of allocation accuracy of inputs. Production of information is costly, so a cost/benefit appraisal is appropriate.

It is vital to ensure that communications convey information clearly and speedily, in suitable form and detail, to the right people. in selecting, procuring, installing and operating an (especially 'high-tech') information system, the interactions with the people involved are vital considerations. Staff needs occur as requirements for the system (capacity, speed, cost, etc.), ability to work with the system and training requirements. System security involves software, data and information as well as hardware. For personal data, compliance with the Data Protection Act, 1984, is imperative.

Questions

1. Explain why any information system should be 'output-driven'.
2. Discuss the importance and consequences of feedback and noise in a communications system.
3. Explain the statement, 'You have the right people but the wrong information system'. Discuss its consequences for staff training and recruitment.

References

Handy, C. B. (1985) *Understanding Organizations* (3rd Edition), Penguin, pp. 356–360
Health and Safety Executive (1983) *Visual Display Units*, London, HMSO

Hodge, B., Fleck, R. A. Jnr and Honess, B. (1984) *Management Information Systems*, Reston, Virginia

Katz, D. and Kahn, R. L., (1977) *An Introductory View of Management*, Harper's College Press, New York, pp. 414–416

Koontz, H., O'Donnell, C. and Weihrich, H. (1984) *Management* (8th Edition), McGraw-Hill

McDonagh, A. M. (1963) *Information Economics and Management Systems*, McGraw-Hill

Simon, H. A. (1960) *The New Science of Management Decision*, Harper and Row.

Chapter 9

Women in construction

In 1988 less than 7 per cent of the full-time construction industry workforce were women (EOC 1988) and over 80 per cent of these are estimated to have been in traditionally women's jobs such as secretarial and clerical (Gale 1991). In 1992 only 40 out of 8452 (CIOB 1992) corporate members of the Chartered Institute of Building (CIOB) were women and in 1991/92 only 13 per cent of university building undergraduates were women (UFC 1992). In the same year only 8 per cent of first-year building undergraduates in polytechnics were women (PCAS 1992).

The Equal Opportunities Commission (EOC) (1990) said in their annual publication *Women and Men in Britain*: 'Women continue to be significantly under-represented in the primary sector (agriculture and energy and water), in most manufacturing, in transport and communications and, in particular, in the construction industry' (p. 13).

The construction industry in Britain defines a large sector of the economy. The size of a construction firm and markets in which it operates strongly influence working practices and the level of professionalization. These are important influences on the gendering of work and the sexual division of labour.

In this chapter a distinction is made between sex and gender. Male domination of the construction industry is described with reference to the workforce, professions and higher education. Theories of sexual division of the labour market are reviewed. Sex discrimination is defined and legislative measures discussed. Some recent and current studies on women in construction are discussed. This field of research is relatively new. Research has been carried out on women in engineering but until now the construction industry has not received special consideration.

The culture of the construction industry is scrutinized. It is argued that the construction industry should feminize if conflict and crisis management are to be reduced. Conclusions identify education as having a key role in the change process advocated in order to challenge the masculine construction culture.

Sex and gender

The terms gender and sex are often used synonymously. This causes confusion and is technically wrong. Sexual division (to do with men and women) and gendering of work (to do with socio-psychological value constructs) have specific meanings. The confusion which is based on a view that biological sex differences explain sexual division of labour is problematical. Any serious discourse on the question of sexual division of labour should be based on a sound theoretical foundation with respect to biological sex differences and gender roles (Garrett 1987).

The theoretical position taken in this chapter is that gender roles are developed as children become adults – psychological theories of the socialization of gender roles are supported. All three main psychological theories of gender development (cognitive development theory, social learning theory and psychoanalytic theory) accept the concept that girls identify with the feminine model and boys identify with the masculine model. These theories agree that parents are the major influence on a child. The theories differ in their interpretation of the nature and particularly timing of this influence. The biological determinists believe that biological factors are very influential in shaping the gender roles of men and women in society. This view is not supported in this chapter. Feminists have done much research which effectively challenges the biological determinist position (Oakley 1972, 1981a). It is relevant to note that psychological theories of gender development are the basis of all mainstream action research and theoretical research conducted in recent years in the domain of women in science and technology.

The construction industry

The industrial and occupational distribution of women is such that women are concentrated predominantly in a few occupations, mostly in the service sector, such as wholesale, retail distribution, hotel and catering. The working conditions and practices in these sectors are traditionally poor, with low wages and short-term employment. This has remained relatively consistent over the last 20 years (Martin and Roberts 1984). Given the generally perceived long-term trend of recruitment problems in the construction industry the importance of attracting more women to construction makes economic sense apart from the fundamental issues of women's rights and equitable status in society.

The cyclical nature of construction output due to its sensitivity to the underlying economic cycle means that the question of recruitment goes on and off the agenda and so too does the question of recruiting a higher proportion of women. The Second World War saw a massive increase in the participation of women in the labour market, including the construction industry sector of the economy. The universally-applied 'marriage bar' in 'better jobs' prior to the Second World War was dropped during the war to enable married women to work and was the single most important effect on gender relations (Walby 1986).

Female employment statistics

The Equal Opportunities Commission (1990) states that in 1988 43 per cent of all employees in employment in Great Britain were women. Over-representation occurs in certain industry sectors: distribution, hotel and catering, repair, banking, finance, insurance and other service industries. Under-representation occurs in construction, transport and all other industry sectors. When considering occupational groups twelve out of sixteen show under-representation by women. In 1988 (EOC 1990)

Table 9.1 Industrial distribution of female employment in the UK

Industry	Female percentage of workforce	
	Full-time	Part-time
Agriculture, forestry and fishing	16.2	9.2
Coal extraction and solid fuels	3.3	1.0
Electricity and gas	17.1	4.5
Metal manufacturing	10.1	1.6
Chemical engineering	25.7	4.0
Mechanical engineering	12.3	3.5
Electrical and electronic engineering	27.4	3.8
Motor vehicles and parts manufacture	10.9	0.9
Other transport equipment manufacture	10.6	1.1
Instrument engineering	23.6	6.6
Food, drink and tobacco	26.3	14.2
Textiles	41.4	7.0
Footwear and clothing	64.0	9.1
Paper, printing and publishing	27.1	7.2
Rubber and plastics	23.7	6.3
Other manufacturing	27.7	15.7
Construction	6.7	5.3
Wholesale distribution	22.7	9.9
Retail distribution	25.3	37.4
Hotels and catering	20.8	46.1
Transport and communication	15.9	4.8
Banking, finance and insurance	36.0	13.0
Public administration and defence	29.7	15.3
Other services	30.8	38.0
All industries and services	25.2	19.8

(*Source:* EOC, 1988)

women represented only 26 per cent of managerial occupations. The EOC says that segregation of the sexes within industries remains an area of considerable concern and argues that women are significantly under-represented in the primary sector, in most manufacturing and in particular in the construction industry.

The construction industry is demonstrably male (Gale 1991, 1992a). In terms of the horizontal sex segregation of the labour market women represented 6.7 per cent (EOC 1988) of the full time workforce in 1981. No reliable vertical segregation data are gathered. The Construction Industry Training Board (CITB) do not gather data on the construction industry workforce by sex. However, an analysis of the 1981 General Census data shows that only 8.4 per cent (Rainbird 1989) of those women in the construction industry occupy managerial positions; just over 82 per cent of them were employed in secretarial or clerical jobs (Table 9.2).

The low proportion of women is reflected in higher education figures (Beacock and Pearson 1989). The percentage of female building undergraduates at British universities (not including old polytechnics) rose from 9 per cent in 1985/86 to 13 per cent in 1990/91 (Table 9.3). Polytechnic entrants to building degrees (now

Table 9.2 Vertical occupational segregation of the construction industry by sex

Occupation	Employment (%)					
	1971			1981		
	Male	Female	Total	Male	Female	Total
Managers	6.9	5.1	6.8	8.7	8.4	8.7
Technicians	4.2	0.6	4.0	5.0	1.3	4.8
Secretarial and similar	3.2	88.6	8.2	2.8	82.2	9.1
Supervisory	4.8	1.6	4.6	4.9	1.7	4.6
Crafts	51.7	1.1	48.8	54.2	3.0	50.1
Operatives	15.6	2.3	14.8	15.0	2.5	14.0
Others	13.6	0.7	12.8	9.4	0.9	8.7
Total employed	94.2	5.8	100.0	92.0	8.0	100.0

(Source: Gale (1991) after Rainbird (1989) General Census Data 1971 and 1981)

Table 9.3 Proportions of female home undergraduate students by subject in British Universities 1985–1991

Session	Female students (as percentage of all students) reading:					
	All subjects	Arch.	Building	Business studies	Civil eng.	Mech. eng.
1985/86	42	27	9	42	11	6
1986/87	43	29	12	43	12	7
1987/88	43	30	12	44	13	8
1988/89	44	31	12	43	14	9
1989/90	45	33	13	43	14	10
1990/91	45	33	13	43	15	11

(Source: UGC (1986, 1987, 1988, 1989) & UFC (1990, 1991, 1992) Table 5)

new universities) showed a small rise from 7 per cent in 1985/86 to 8 per cent in 1990/91, with a high of 10 per cent in 1988/89. Statistics given by the Polytechnics Central Admissions System (PCAS) show only application and admission rates (Table 9.4). According to a recent CNAA study (Beacock and Pearson 1989) there is a greater proportion of female students in built environment subjects at universities than at polytechnics. Architecture, building services engineering and building surveying have shown increases in female undergraduates. This could be disputed in considering building.

Also, the professions demonstrate the male characteristics of the industry with 40 women out of 8,452 corporate members of the Chartered Institute of Building (CIOB) in 1992 and 594 women out of a total of 32,562 in all grades, representing less than 2 per cent women (CIOB 1992). It should be said that the CIOB is attempting to address this situation, a point demonstrated by the fact that these data are a function of the Institute's self-critical review. A Women in Building Consultative Committee

Table 9.4 Proportions of female home undergraduate students by subject in British Polytechnics 1985–1991

Session	Female students (as percentage of all students) reading:					
	All subjects	Arch.	Building	Business studies	Civil eng.	Mech. eng.
1985/86	42	26	7	47	8	5
1986/87	42	24	9	46	5	6
1987/88	44	25	7	48	9	6
1988/89	47	24	10	50	13	6
1989/90	46	23	11	50	12	6
1990/91	46	20	8	50	13	7
1991/92	49	22	9	49	12	6

(Source: PCAS (1987, 1988, 1989, 1990, 1991, 1992, 1993) Table C2)

Table 9.5 Membership of some construction industry professional bodies by sex in 1992

Professional bodies	Membership*					
	All grades		Corporate		Fellows	
	Male	Female	Male	Female	Male	Female
Chartered Institute of Building	31,968 (98.2)	594 (1.8)	8412 (99.5)	40 (0.5)	2280 (99.9)	2 (0.1)
Royal Institution of Chartered Surveyors†	83,555 (92.7)	6575 (7.3)	37,638 (92.5)	3043 (7.5)	26,141 (99.4)	159 (0.6)
Institution of Civil Engineers	76,423 (96.8)	2497 (3.2)	43,027 (99.0)	416 (1.0)	6174 (99.9)	5 (0.1)

(Source: The membership departments of each institution)

* Percentages in parentheses
† RICS Corporate members are called Associates; Fellows include figures for Members and Fellows

meets annually to discuss this question. Women account for less than 4 per cent of the Institution of Civil Engineers (ICE) and 3.2 per cent of the Royal Institution of Chartered Surveyors (RICS) (Quantity Surveying Division) (Greed, 1991). Table 9.5 gives figures for members at all grades for all divisions of the RICS and statistics for the CIOB and the ICE.

Women in training and education

The Engineering Industry Training Board (EITB) (now the Engineering Construction Industry Training Board) has promoted several initiatives over the last 20 years. Peacock and Eaton (1987) say: 'The Engineering Industry Training Board discovered women as long ago as 1975.' The CITB did purchase a number of places on two EITB 'insight courses' in the early 1980s. In the late 1980s the CITB collaborated with pilot insight courses (Gale 1989, Baker and Gale 1990) and later developed experimental insight courses for school students (Gale 1992b).

In an article published in 1987 the senior careers adviser for the CITB explained that there were in 1986/87 three women only on Building Technician Certificate courses in Britain.

There were 53,000 employers registered with the CITB and 20,000 companies involved in their training scheme. These courses lead to trade jobs in the industry. It would appear then that the industry is fishing in only half an ocean and not recruiting women.

The current growing interest in attracting women has been seen by some teachers, parents, school students and careers advisers as opportunistic. In other words: does the construction industry want to encourage more women or is it just seeking them as a last resort? Further, the industry has a poor public image and research shows that teachers, parents, careers advisors and school students have, at best, only a vague, superficial knowledge of the industry (Baker and Gale 1990). Due to the variety of courses and diversity of career paths available in the construction sector even dedicated professional careers advisers find the subject of careers advice highly confusing.

Recently the CITB has made progress in the area of careers advice for professional careers in the construction industry. The Construction Careers Service (CCS), a part of the CITB, has a number of initiatives and provisions to improve the liaison with schools and careers services. These include work shadowing, provision of speakers at careers events, free brochures, videos and literature, careers seminars for teachers and careers advisers, schools/industry links and curriculum centre (CITB 1993). The importance of making these initiatives 'gender-inclusive' is being taken seriously by the CITB. The initiatives all attempt to introduce a positive attitude towards the entry of women.

Cockburn (1987) cites the words of a careers adviser talking about the sexual prejudice surrounding Youth Training Schemes (YTS) in the CITB:

> Everyone knows the CITB shows a lot of prejudice towards girls but the Manpower Services Commission (MSC) – and this would have to be at national level – would not take the CITB to task because it can't afford for the CITB to pull out of the scheme. So really they condone it. . . . And I mean, government aren't really interested in equal opportunities (p. 64).

The early lack of work and research in the construction industry makes the work and research of the EITB and research associated with its initiatives worthy of investigation.

International comparisons

Wells (1990) states:

> The construction labour market is clearly segmented along sexual lines. In all countries skilled and supervisory tasks are undertaken exclusively by men. In

some countries women undertake clerical jobs; in others they are employed as unskilled labourers, helping out in the most menial of tasks. While in many countries (from India to the USA) educated women are beginning to break into higher professional jobs of architect, engineer, surveyor, there is still considerable resistance to the entry of women into the vast mass of middle level, skilled and semi-skilled, jobs in the construction trades. In European countries such resistance would appear to stem from deeply ingrained prejudices on the part of employers and other (male) workers, about the suitability of women for construction work (p. 28).

From these statements it is apparent that the problem of women's role in the construction industry is an international one. However, cultural differences are a very important influence on the actual functions undertaken of women in the construction process. What is common is the status of women in the construction sector: one of being controlled by men. This position is explained by the concept of patriarchy in which males control (Hearn 1990).

Comparative statistics for three western countries, UK, France and the USA, show that women are under-represented comparably in all three countries (Tables 9.6 and 9.7) (Dale and Glover 1990).

The sexual division of labour

Discrimination occurs in the labour market and can be demonstrated through a study of differentials in earnings for identifiable minority groupings: women and ethnic minorities. It is also possible to identify discrimination through what is called occupational segregation in the labour force.

It is possible in theory to distinguish between those differences which are a consequence of:

- attitudes
- differences in occupational level
- differences in pay
- participation in the labour force

Some attitudes are acquired before entry to the labour force. In the case of differences in occupational level this is called occupational discrimination. Differences in pay is termed wage or income discrimination. There is also participation discrimination.

Men and women are equal in their capabilities in almost every respect, provided that women are given the same opportunities for training, employment and promotion as men. It is also assumed that should men come to accept responsibility for raising the young, caring for the sick and elderly, and looking after the home, the participation of women in the labour force would be essentially equal to that of men. Tsuchigane and Dodge (1974) p. 5.

Economic theories of sex discrimination are hybrid offshoots of theories of racial discrimination. This is unsound because of a number of differences between the concepts of sex and race.

Table 9.6 Distribution of women and men in building and civil engineering in 1983/4

Industry	Employment distribution (%)					
	UK		France		USA	
	Male	Female	Male	Female	Male	Female
Agriculture	3.5	1.3	8.7	7.4	4.5	0.8
Energy and water	4.8	1.1	1.8	0.6	2.3	0.7
Extraction and processing etc.	4.9	2.0	5.0	2.0	3.6	2.3
Other manufact.	15.2	5.6	14.0	5.8	15.3	7.2
Building and civil engineering	12.0	1.4	12.4	1.6	8.7	1.1
Distributive and hotels, catering	15.4	26.7	16.1	18.6	19.1	20.3
Transport and communications	8.7	2.9	7.6	3.7	4.7	4.9
Banking and finance	6.9	9.1	6.5	9.0	8.9	12.7
Other services	17.9	40.1	18.3	41.2	25.5	42.9

(Source: Dale and Glover (1990))

Table 9.7 Female ratio of segregation by sex in France, UK and USA

Industry	Ratio of percentage of women in industries to percentage of total women in employment		
	UK	France	USA
Agriculture	0.5	0.9	0.3
Energy and water	0.3	0.5	0.5
Extraction and processing etc.	0.5	0.5	0.8
Other manufact.	0.9	1.0	1.0
Building and civil engineering	0.2	0.2	0.2
Distributive and hotels, catering	1.3	1.1	1.0
Transport and communications	0.5	0.6	1.0
Banking and finance	1.2	1.2	1.2
Other services	1.5	1.5	1.3

(Source: Dale and Glover (1990))

For a start there is the issue of family roles with respect to women. Also the idea that women and men are essentially different societies is demonstrably not the case. Sloane (1985) argues that research shows there to be a difference between sexes in terms of achievement-related motivation but no difference between black and white people. Women seek better interpersonal relations and working conditions compared to men and men seem more highly motivated by opportunities for independence and upward mobility (Carline, D. *et al.* 1985). Gale (1993) challenges the notion that women in the construction industry are motivated differently from men. Earlier motivation theories were gender-blind according to Dex (1985).

In considering the labour market, Hakim (1979) explains that it is horizontally and vertically segregated occupationally. Horizontal segregation based on sex occurs where men and women are found to carry out different types of job or occupation. Hence, there are male-dominated industries and professions. According to Hakim (1979) horizontal segregation is declining.

Since about 1900 women have made up approximately one third of the labour force and vertical segregation has been intensifying (Hakim, 1979). Vertical segregation gives rise to what can be termed primary and secondary sectors of the labour market. The former refers to managerial and technical jobs and the latter to lower grade manual unskilled jobs. The primary sector has the characteristics of security and relatively high pay whilst the secondary sector displays characteristics of lower pay and insecurity (Barron and Norris, 1976). Hakim (1979) argues that education plays an important role in this occupational segregation. This concept is called the dual market. The dual market reflects socio-economic, political and technological differences between cultures and individual differences in general. This explains why in other cultures occupational segregation may not be based on gender. For example, in India the dual labour market is said to be based primarily on racial and ethnic differences or rural and urban origins. In some countries, for example Latin America, it can be shown that women were the first professional grades (Barron and Norris, 1976). They go on to say that changes in the dual labour market can be brought about by either eliminating the dual nature of the market, by restructuring it or by changing the basis of discrimination. The reality of contemplating such changes would inevitably be very long term and they make no recommendations for how these changes could be achieved.

Types of discrimination

Some economic models of discrimination consider the stereotyping caused by family role reinforcement to be significant and slow to change. It is in fact this stereotyping in family role models that distinguishes sex discrimination from racial discrimination according to Chiplin and Sloane (1976). They put forward a sophisticated economic model of discrimination. They discuss pre- and post-entry discrimination. Pre-entry discrimination is said to occur due to differences in educational opportunities between girls and boys and through overt discrimination such as not hiring or promoting due to sexual discrimination. Post-entry discrimination occurs within organizations and can be termed institutionalized discrimination. It pervades the social culture of the organization.

Madden (1973) describes three types of discrimination:

1. occupational
2. cumulative
3. wage

Wage and occupational discrimination have been mentioned previously. Cumulative discrimination is based on the concept that the labour market currently displays discriminatory elements caused and sustained by the impact of previous discrimination. This can be used to explain women's inferior role in the economy. Meehan (1985), in comparing United States and UK legislation, categorizes economic explanations for employment patterns into three views which, she says, overlap somewhat.

- Sociological approaches or supply side analyses.
- The interaction between pay differentials and occupational distribution of women and men.
- Dual market approaches – primary and secondary sectors.

In the case of supply side analyses, Meehan demonstrates that prior to equal opportunities legislation in both the USA and UK women were not being educated in scientific or industry-orientated subjects. The theory says that industry increases profits through discrimination because employers do not employ women in jobs for which the firms would have to invest in training. Some explanations concentrate on the idea that employers having a 'taste' for discrimination causes low pay for women, and employees having a 'taste' for segregation causes low pay for women. This approach says that profits are reduced in order to 'enjoy' discrimination.

Imposed segregation

Dual or segmented labour markets are based on easily screened characteristics such as race, gender and educational qualifications. Meehan explains that a vicious circle operates by which workers take on the characteristics that go with the jobs in the secondary sector. These include absenteeism, high turnover and an apparent lack of interest in acquiring skills prior to and after entering the labour market.

Walby (1988) argues that segregation in employment is the key to the wage gap between men and women. Holton (1989) states that women's salaries have improved in recent years but despite the Equal Pay Act (SDA) 1975 women today still earn on average less than men. Women's earnings in 1971 were less than 65 per cent of men's. By 1976 they had risen to 75 per cent but since then they have risen no higher. She reports on a survey carried out by Staffordshire Polytechnic on graduate pay. This revealed that women graduates tended to earn less than their male counterparts and had lower status jobs three years after entering the labour market. The salary expectations of women graduates were far lower than males.

Traditionally women's position in the family has been the basis for explanations of the disadvantages women face in paid employment. There is a debate between theories based on material factors and those that emphasize the cultural process. Also a more specialized debate on the theory and concept of patriarchy and the position of women in society. Walby (1986) argues gender inequality in the labour market is a significant determinant of women's position in the family and vice versa. Dex (1987) suggests that for women there is a more complex structure of market segmentation. She says that there is a special full-time work primary sector for women. Education is an important aspect of socialization and gendering.

The EOC (1990) states that the fact that boys and girls take different subjects at school influences their future career paths. There is a demonstrable distinction between boys and girls and the subjects they take at school. Almost two-thirds of pupils who attempted A level biology in 1987–88 were girls when around 70 per cent of those attempting physics were boys. In England in 1987–88 twice as many girls as boys attempted English at A level and twice as many boys as girls attempted mathematics.

Orr (1985) says that since the early 1970s equal opportunities for both boys and girls in schools have been implemented largely through equal access policies associated with the SDA 1975. He argues that in the majority of schools many teachers (particularly males) are not interested in equal opportunities and argues for training to help teachers understand the subtle processes that lead to undue differentiation by sex among boys and girls during school years. Orr concludes that over ten years initiatives to reduce sex differences in schools have increased but questions whether this increase will lead to a major shift in the aspirations and expectations of boys and girls in schools. Srivastava and Fryer (1991) argue that in patriarchal western society the sexual inequalities in work and social institutions are caused by schools, parents the media through the socialization process by which girls are socialized differently from boys.

The Equal Opportunities Commission (1986, 1987a, 1987b) states that inequality between sexes results in women:

- being concentrated in only a few occupations
- receiving less opportunities to learn skills
- receiving lower pay

The Commission argues that occupational segregation should be eliminated because:

- there is a waste of human resources
- it causes a lowering of morale in the workforce and a souring of industrial relations likely to lead to industrial action
- it causes unlawful discrimination
- it leads to breaches in the Equal Pay Act (EPA) 1970

Hakim (1979) argues that there is a relationship between the work profile of women and the fact that women are concentrated in the secondary market. Also she identifies the importance that education plays in this occupational segregation. Work patterns are, it seems, linked to family roles and relationships and personal characteristics.

It is said that because women tend to have 'career breaks' associated with child rearing this can be connected with their place in the labour market (Manley and Sawbridge 1980). However, there are significant differences in vertical and horizontal occupational segregation between organizations and the nation as a whole. Hakim (1979) distinguishes between *de fine* or direct and *de facto* or indirect mechanisms causing occupational segregation.

Discrimination in job advertisements and interviews
The Equal Opportunities Commission (EOC 1987b) have published their views on the causes of job segregation in detail and list these under a number of headings.

Under the heading of advertisements it identifies a number of contributory factors. These include explicit statements that women are excluded from consideration. Implicit statements which imply that women are excluded such as words that are gender-specific. Also, statements which imply that women would not be interested. The EOC suggests that if neutral language is not possible then a statement should be made clarifying that women and men should both apply. Illustrating a preference exists in testimonials by men only or photographic and other images with a male slant and the placement of advertisements in journals read only by men or likely to be read mainly by men.

Under the heading 'selection and interviews', they say that there are many assumptions which contribute to women being discriminated against. These include: women not being fit or able to carry out dirty hot work; women displaying more manual dexterity than men; women being better at coping with monotony; men preferring to take jobs requiring initiative whereas women tend to decline; women are said not to be willing to take on leadership and responsibility (this assumption particularly hampers women's ability to gain higher level industrial and professional positions); management anticipates reluctance of men to work with women where there is currently a high order of segregation excluding women; women's ability to work long hours and other assumptions concerning strength and stamina associated with working hours; a number of misconceptions about women's physical strength and its relevance to working requirements; objections to one particular sex.

Surveys have shown a general lack of consistency in judging men and women due to bias against women.

Other headings include: job agencies and their engagements, waiting lists and practices of employment, recruitment practice and school careers services and their role in informing and participating in the placement of school leavers in jobs.

Women's action groups

The Women In Construction Advisory Group (WICAG) set up in 1984 (WICAG undated) has the aim: 'To improve the training and employment position of women in the construction industry, in consultation with employers, trade unions, policy makers, training bodies and women working and training within it.'

WICAG published a guide in the late 1980s for construction employers on the recruitment and employment of women. This gives advice on preparing an equal opportunities policy, recruitment and employment of women. It includes checklists on related topics. According to *Women & Manual Trades* (WAMT 1993) the London Boroughs Grants Unit (LBGU) withdrew all funding from WICAG in the same month it stopped funding Matrix women's architect scheme.

Factors causing marginalization of women

Spencer and Podmore (1987) list ten factors which contribute to the professional marginalization of women in male-dominated professions.

- Stereotypes about women.
- Stereotypes about the nature of professions and professionals.

- The sponsorship system.
- Lack of role models and peers.
- Informal relationships.
- The concept of professional commitment.
- The unplanned nature of many women's careers.
- Women's work.
- Client expectations.
- Fear of competition.

Cooper and Davidson (1984) say that because line managers, including the more senior, tend to graduate from the shop floor, the sexual emphasis extends to the top, stemming not simply from suitability but from male pride. They argue that it is the very style of management regarded by many men as essential to their industry that is endangering it.

According to Carter and Kirkup (1990), who interviewed women engineers, quite a few thought that women had special attributes to bring to the engineering industry and that they were better than men in certain aspects. This involved the notion that women were better at people management. The women engineers they interviewed did not want to be characterized as 'women's libbers' and they did not want special consideration because of their sex. Kirkup and Carter argue that these female engineers took advantage of the changes brought about by feminism in the last twenty years but few of those interviewed self-identified as feminists. Men were seen as very important as role models, mentors and providers of crucial information. The women engineers had strong individual identities but did not see themselves in a category called women. They quite often denied being actively discriminated against.

Cockburn (1985) explains that industrialists generally express the view that there are willing managers but absent women. From her research she found there to be a perceived need by employers to reach into the schools; that it was the teachers', parents' and careers advisers' fault and fundamentally it was women's fault. She argues that work as well as family and schooling is a gendering process and that 'the social process of gender construction and formulation of gender difference are important mechanisms in sustaining male dominance. It is an expression of men's hegemonic ideology, and one into which women get swept.'

Cockburn (1985) argues that 'equal opportunities' is not in itself enough and that the sexual division of labour has developed over a long time and been maintained by the gendering process. She makes the case for a more radical perspective than has yet been attempted by the Equal Opportunities Commission or the Engineering Council.

Sex discrimination legislation

The law defines sex discrimination as both direct and indirect. The SDA 1975 employs the concept of taking a man as the standard or 'norm' and providing that a woman with similar characteristics or similar standing is treated on the grounds of her sex less favourably than a man would be in those circumstances, this is discrimination which is unlawful (Walker 1975). This definition is very important because if what might at first sight appear to be sex discrimination is not specifically covered by the SDA then it is not unlawful. There is, of course, discrimination of some sort in every choice made. Therefore if criteria used in an interview are not based on those which would be considered deliberately made to ensure a man is selected, the discrimination concerned in such an interview would not contravene the SDA.

The SDA applies equally to men as it does to women.

Another point worthy of consideration is the fact that it is the effect or essence of certain behaviour or actions which must be judged, not the intentions. Direct discrimination occurs when an employer has a preference for a member of one sex and denies employment to an applicant of the other sex on grounds of sex alone or differentiates between their employees in access to training, promotion and so forth for the same reason.

Direct discrimination also occurs when different standards or requirements are demanded of men compared with women.

Indirect discrimination

Indirect discrimination occurs when requirements or conditions are applied equally for men and women but where there is a considerably smaller population of one sex which is able to meet the requirement and this cannot be justified in terms of the requirements of the job. For example, minimum height restrictions where height is irrelevant to performance.

Legal bodies in the UK and USA

The UK SDA 1975 provides for the Equal Opportunities Commission (EOC). This has authority to require offenders to cease and desist subject to convincing a court that an injunction should be issued. In the USA Title VII of the Civil Rights Act 1964, Executive Orders 11478 of 1969 and the Equal Opportunities Act 1972 provide for the Equal Employment Opportunities Commission (EEOC). The EEOC does not have the same 'cease and desist' authority as the UK, EOC being required to pass litigation to the Department of Justice. But its recommendations for legal action may have been more wide-reaching because of the US concept of class action. Class action is a great deal more difficult to argue in the UK. However, since 1979, it has also been somewhat restricted in the USA due to more narrow interpretations of what constitute the common characteristics of a class by the judiciary. There is no doubt that disobedience is a great deal more costly in the USA than in the UK (Meehan 1985).

Legislation in both the UK and USA does allow for sex discrimination as a 'genuine' occupational qualification in certain cases.

Affirmative action

According to Walker (1975) positive discrimination is allowable under the 1975 SDA if there is a demonstrably low proportion in the UK as a whole or a region: this is under Section 47. Also under Section 48 positive discrimination is encouraged if there have been very few or no women recruited in the previous twelve months.

This can be achieved either by encouraging women to take the particular work function in question or by training women only to do such work. Also universities can apply for exemption from the normal 'equal treatment' in admissions to allow women (or men) to undertake training in disciplines for which they are demonstrably absent.

Section 49 of the SDA says that it is not unlawful for elected bodies to make special arrangement to reserve seats, make extra seats available or create minimum quota for women. It defines these bodies in the Act under Section 12 subsection 1 as 'An organization of workers, an organization of employees or any other organization whose members carry on a particular profession or trade for the purposes of which the organization exists.' These are, for example, trade unions, professional associations or employers' organizations (Walker 1975).

Recent success in introducing a 'model contract for building' was reported at a conference in January 1993 held by Women & Manual Trades (WAMT 1993), funded by the European Social Fund. Due to the leverage given to the client of a £2.9 million project to restore an historical women's school in Liverpool, Liverpool City Challenge has now adopted the principles of a model contract for all of their building works. The model contract states that 'the employer is committed to positive action and requires contractors and sub-contractors to demonstrate their commitment to the recruitment, training and employment of women and black people and people with disabilities'. The model contract states: 'The contractor shall guarantee to the employer that certain specified language or behaviour would not be permitted to take place on site nor that any printed material be brought to site.' It then goes on to list in more detail requirements under the headings of language, behaviour and printed matter.

Some studies on women in construction

Srivastava (Srivastava and Fryer (1991) and Srivastava (1992)) is carrying out an investigation into access programmes in which she is attempting to identify specific barriers to women entering construction-related higher education and the nature of the image and culture of the construction industry portrayed by schools, careers advisers, teachers and parents and the role of higher education in enhancing the recruitment of women into the construction industry. She argues that equal opportunities policies are not always translated into practice with lecturing staff often indicating a lack of understanding and a degree of apprehension towards widening access. She says that construction higher education was one of the last areas to respond to opportunities for widening access and says this problem is endemic to construction departments throughout higher education.

She states that female construction students and women construction professionals revealed traditional attitudes towards male and female roles, abilities, aspirations which result in women being steered away from construction even when they show an early interest in the industry. She explains that women are not given the 'right' experience and they often lack confidence to apply for courses for which they are qualified.

Women often do not want a 'traditional female' career but advice given frequently omits to mention construction and related areas of work and study. She says that female students feel excluded by terminology and language of tutors. These points are all supported by the work carried out and published by Farish (1987) in which she cites specific responses to the question asked of built environment academics. Farish cites Spender (1985):

> For centuries men have been encoding knowledge about themselves in which they and their concerns were central; which they took for granted as reliable knowledge, and which confirmed their existence and their supremacy in the world (p. 138).

West (1982) has carried out doctoral research into career choice and occupational perception of professional builders and quantity surveyors. She concludes that although many interviewed had drifted into construction careers they would choose the same careers if they had their time again. She argues that professional careers in construction should be promoted as worthwhile and satisfying but that the image of the professions must be improved by promoting the intrinsic satisfaction in these occupations. She states that the most significant conclusion is in the area of occupational stereotyping and work role self-perception of professional builders and quantity surveyors. She says that members of both professions appear to fail to identify their work role with their stereotypes of their professions. This suggests that the entrants to construction industry professions of builder and quantity surveyor choose their professions independently of the matching process of occupational stereotypes and self-perception. She argues finally that career guidance must be improved before limiting educational subjects are chosen at school. Her final conclusion does not relate to the current situation in that most quantity surveying and building degree courses accept a very wide set of entry requirements with respect to A level subject passes and alternative entry qualifications.

Wilkinson (1992a, 1992b) is carrying out research on graduate employment and higher education courses in civil engineering. She has conducted a survey of undergraduates. She found that the image of the construction industry, career planning, occupational status and lack of perceived professionalism in construction were features of the construction industry professions. She argues that an earlier study undertaken by the Institution of Civil Engineers reached similar conclusions.

Wilkinson (1992b) reports on a survey in which she questioned contractors, consultants and local authorities attempting to investigate attitudes to the employment of women, child care and women's career paths in the construction industry. She reports that a low proportion of employers questioned had an overtly discriminatory attitude towards women. However about 20 per cent thought women a distraction on site and that supervision was a 'man's job'. She argues that employers differentiate between various different construction process functions with respect to the suitability of women workers. She says that this could influence the difference in career paths of men and women in construction. She concludes that the prospects for change are not 'grim' as employers do not appear in general to be inclined to refuse employment to women on principle (see also CITB 1988).

Rodgers (1991) investigated the perceptions of the construction industry held by A-level students, female undergraduate civil engineering students and female civil engineering graduates in Northern Ireland. She argues that girls are not encouraged to enter the construction industry because of lack of knowledge in schools and because of the poor image of the industry held by the general public. She says that although female and male students have a good basic knowledge of engineering they do not have the specific knowledge required for a civil engineering career or the construction professions in general. This she attributes to inadequate careers advice in schools and she states that the problem is with the careers advisers themselves and their lack of objective knowledge about the construction industry and its professional careers.

She reports that female graduates said the variety of work and levels of satisfaction in their careers to date were the most appealing aspects of their jobs. She says that there is no reason why these findings should not be similar for the British mainland but does not substantiate this claim.

Pearson (1991), a female construction company director, carried out research into why there are not more women taking up construction work in the construction trades, considering the position taken by the CITB, trade unions, employers, school leavers, parents and teachers. In her conclusions she finds that there is no physical reason why women should not do manual work in the construction industry. She considers the historical role of women in the economy and the nature of the heavy physical work carried out by women in developing countries.

She says that female school students do not find the construction industry attractive, finding it heavy and unfeminine. She argues that employers discriminate against women in the construction industry and even in the case of those employers who professed to be in favour of employing women none of them actually employed a single women in a trade or manual position. She reports that the Building Employers Confederation (BEC) says that there is a need to be realistic with regard to women's physical ability and that the role of women should be concentrated on crafts such as painting and decorating, requiring manual dexterity which is a blatant stereotype-based position (cf EOC 1987a and b). She argues that approaches to increasing the attractiveness of working in the industry for women returners such as career breaks, creche facilities, part-time positions, shared jobs, term time working and flexible hours were all difficult to achieve due to the nature of construction work. She does argue however that an increase in women in the industry might result in improved health and safety, working conditions and an increase in the available pool of labour for the industry.

Greed (1991) has carried out a study of the surveying subculture. She presents an important conceptual model for the investigation of women in the context of the built environment. She says that her study was concerned with the nature of surveying subcultural values and processes. She relates these processes to the spatial aspects of our built environment and argues that the educational process in surveying is a gatekeeper to the male-dominated surveying profession.

She says that the educational process teaches women to fit in and maintains the subculture of surveying. Men genuinely want to increase the number of women in the surveying profession. However, they do not know how to treat women when they are there. Typically males are gallant and treat women as honorary males.

In the state of Victoria in Australia, Pyke (1991) has carried out research on women in manual trades in Australia. She measured the current participation and acceptance of women employed in the building industry. Evaluated employer attitudes towards women's employment in the building industry and investigated strategies to assist women in gaining access to employment within the building industry.

Her draft interim findings state that there is a strong stereotyping of women in terms of their perceived limited ability to perform as effective building workers. She suggests that a small proportion of progressive employers do not take this position and says that women can not be seen in such narrow terms.

She argues that there is little willingness on the part of employers to support an increase in the employment of women in the building industry. These attitudes centre on a widely-held belief that women are not physically capable of carrying out building work and that negative male attitudes are the major problem faced by women in the industry.

Some interesting responses were received in answer to a question about advantages of women in the building industry. Although 50 per cent either did not respond to this question or said there were no advantages, 35 per cent answered positively with a number of interesting ideas. These included the likelihood that the workplace would be more harmonious and balanced and would lead to a less sexist work environment and promote equal opportunities. It was said by some that the behaviour of men in the industry would be improved and the presence of women would help to change male attitudes. Women were seen as more efficient and conscientious. There were some who thought the industry's image would be improved by employing more women.

Employers were generally quick to recognize that women faced problems, negative attitudes, bad language, lack of acceptance by male co-workers, sexual and general harassment and intimidation, discrimination and prejudice, lack of power and difficulty in being allowed to fit in.

Research commissioned by the CITB, published in 1988, investigated the image of the construction industry and specifically investigated women in construction. The research set out to examine the influences on career choice, the Youth Training Scheme (YTS), the work patterns of young people, the image of the construction industry and the level of interest in training for jobs in the construction industry. 9208 young people and parents were interviewed in 12 areas of England and Wales. These areas included those with a high Asian and Caribbean ethnicity.

Career ambitions of those interviewed mentioned relatively few construction occupations. Many appeared to start thinking of career alternatives at about twelve years of age onwards, coming to a decision at around 15 and 16 years of age. Parents were the single strongest influence on career choice amongst all groups interviewed. Careers advisers and teachers were also important. Main attractions for careers in general were security and skills acquisition. The construction industry was not seen as well respected and was also considered not to be for non-whites or women.

Images held of the industry were not very clear. There seemed little understanding that the construction industry involved any more than bricklaying, concreting and plastering. Positive images centred on pay and skills. Negative aspects were cited as danger and dirtiness. The report found that there was a widely held view that women would not be treated equally and would have little chance of getting on and that even if they did they would not get a good job and would face sexual harassment at work.

Females found the industry more attractive when it was linked with socially-useful aspects of life such as the repair of old people's accommodation and the building of schools.

Over 60 per cent of both females aged 18–24 years and their parents thought that the best jobs in the construction industry would go to men.

Telephone interviews were carried out with 1003 employers in the construction industry in England and Wales. Whilst the majority (76 per cent) of employers recognized that there are many in the industry who would not employ women they are less likely to agree that women are able to do many of the jobs satisfactorily (61 per cent) and less likely to perceive their clients as having no confidence in women (37 per cent).

In response to the statement: 'There are few jobs in the industry which women can do satisfactorily' 61 per cent agreed and 27 per cent strongly agreed. The larger companies appeared less prejudiced. The statement: 'There are a lot of employers in the industry who would not employ women' yielded overwhelming agreement.

Gale (1993) finds males and females appeared to converge in their images of construction from school through university and into employment in the construction industry. Although both sexes share common images of the construction industry their behaviours may differ. This is extremely difficult to ascertain.

He finds that both sexes appear to have the same level of knowledge of the construction industry but there is no evidence that more knowledge discourages either sex from entering the industry. In fact, lack of knowledge could well be a discouraging factor for women. Knowledge was seen as extremely important by school students considering a degree in construction. Careers teachers and careers advisers were perceived by school students, undergraduates and graduates to provide inaccurate and inadequate information on the construction industry.

Early childhood events and experiences strongly influence the potential for a person to be influenced or changed later as a school student. Information and its context are also gendered.

A letter to a London higher education institution department of building studies illustrates the nature of sexism in the industry. A large contractor sent the letter to the work placement tutor stating:

We are interested in taking students from March 1988 to September 1988 however our choice is initially from young men who are either from Wales or who are prepared to eventually settle here if full time employment is found for them. We would be obliged if you would circulate amongst your students our Company name and we would be happy to receive applications from them. We enclose 10 application forms for distribution.

The letter was posted on the notice board for students to read. When the tutor was tackled about the overt sex discrimination of this letter he did not seem to be aware of this fact. When it was pointed out that the legality of the letter was questionable as well as his part in publicizing it he removed it from the notice board and posted a modified text.

The culture of the construction industry

Arguably construction culture is male and inherently conflictual. It would benefit the industry for a more female culture to be developed. This would reduce conflict and attract a more varied intake of people to positions in the industry. This would lead to a change in working practices and management style. As the industry feminized it would be demonstrably less male dominated in the physical and psychological sense; gender inclusive.

It seems that if construction culture needs to be feminized the curriculum must address the question of interpretation of gender differences so as to free up the thinking of construction students.

According to Handy (1985) cultures can not be defined precisely. They can, however, be differentiated and a good fit between an organization's prevailing culture and an individual's cultural preference leads to a satisfying social contract. He explains that masculinity, one of four cultural dimensions, is connected with ambition, the desire to achieve and to earn more, whereas its opposite, femininity, is more concerned with inter-personal relationships, the environment and a sense of service. Men tend to be concerned with quantitative and women qualitative aspects of life. These are stereotypic and whilst possibly demonstrable are not necessarily linked with gender. The differences may be explained by the difference between the material experiences of men and women (Dex, 1988, Cockburn, 1983 and 1985). However, different cultural types can be described and compared. Cultural differences can be defined at different levels: national, industrial, organizational and individual.

Greed (1991) gives many graphic examples of the male surveying culture (a subculture of construction) and how women cope. There are identifiable differences between various subcultures. These may in some part be due to the relative levels of professionalization, historical determinants and prevalent commercial organizational types.

Greed describes negotiation:

> When men meet (based on what men have told me) they are likely to spend a while discussing the weather, cricket, women, cars, etc., and then almost as an afterthought say, 'my goodness look at the time, let's see what I can do for you'. There then follows a prolonged period of competitive discussion in which both sides want to save face and protect their egos. Men tell me that men always like to haggle and there are unwritten rules about offering high unrealistic figures first to protect the pride of each side. Men have always got to have the last word and win, or choose to concede (p. 151).

She goes on to describe the approaches taken by women. They either consciously try to emulate men or are more likely to be more direct in putting their final offer. Also a characteristic of women's negotiation seems to be to put their final offer forward at the beginning of the negotiation. Greed finds that those who have negotiated with women find the experience much more straightforward, with far less posturing and no male egos to protect.

Greed explains that a surveying culture exists in the surveying departments of educational establishments and writes at length about this phenomenon and how women cope with it.

Roger Pauli, Group Managing Director of Stuart Crystal, speaking at The Women's Education Conference (WEC, 1989) made some revealing comments about male culture in British industry. These could certainly apply to the construction industry. The following is compiled from Pauli's speech to a predominantly female audience:

> I am not sure that I believe that men don't know what they are doing when they stereotype women. Our organizational structures have been devised within a male-dominated society. They are primarily command structures. I believe they are to do with how to gain and retain power: how to get people to do what you want them to do. It isn't necessarily how to organize the carrying-out of tasks. I think that is one of the reasons women can feel so uncomfortable in industry. We borrowed these organizational structures from the army – the only organization available when industry started. We are talking mostly about people's thinking directing the activities of others. In a healthy environment, people thought about what they felt and did. In a command structure thinking was done at the top and people lower down in the organization were not expected to think about what they did. How, then could they develop any decent feeling or conscience about what they were doing? We need to help society to recover from this terrible malaise we have got. Hold fast to your courage, stay female, and help us men to find more of the feminine within us (p. 23).

Views of construction students

The construction industry is demonstrably male as has been outlined earlier in this chapter. Not only is the workforce male, the prevailing culture and ethos of the industry appears to be extremely male, characterized by comments like the ones below. These were taken from the responses of males to a questionnaire survey of young construction industry trainee professional managers studying on a part-time basis for professional examinations (Gale 1987):

> The natural male instinct for attraction to women with implied sexual 'innuendo' helps women considerably in their careers.

> The women fundamentally are cheap, docile, unionised clerical labour.

> I don't really understand why women want to work in a traditionally male industry – compare with nursing. (I don't have anything against women.)

These are not just important because they demonstrate how male, almost how misogynist, the construction industry is but because they are young people who, it could be argued, might be expected to hold more enlightened views.

Gale (1992) uses the following key words to sum up the construction industry's culture:

- crisis
- conflict
- masculine

Although much research continues to be done in the area of risk management the underlying tendency in construction management seems to be that of a willingness to engage in crisis management if at all possible. In fact, reduction of risk and uncertainty is an anathema in construction culture. People do not join the industry to have an easy life, they thrive on crisis management.

Conflict is a part of everyday life in the industry. The management of operative labour involves high expressed emotion and the image of the aggressive 'barking' foreman is a generally held one both in and outside of the industry. The handling of conflict is seen as an important management skill at all levels of management. A great deal of space is taken up in the construction press with contractual matters. Conflict is almost assumed at the inception and completion of a contract.

There are a lot of men in construction; so it is obviously male. However, masculine is meant here in a gender value and behavioural sense. The customs and working practices of construction managers and operatives alike seem to be very male. The image is held both in and outside the industry of the hard-drinking, sexist, hard-playing male. To get on, women have to fit into the culture. The mobility of the workforce is particularly high in the construction industry. This contributes to some extent to the exclusivity of the male 'clubiness'.

All of the above characteristics are to a certain extent carried over into the subculture of the further and higher education departments relating through their strong vocational orientation to the construction industry. It is easier for those outside of this subculture but within the educational environment to observe and comment on this tendency.

Construction students share a common orientation towards a career in the industry. Also, because a high proportion of them are, or have been, employed in the industry, they bring the industry's values into their courses and departments. In this way the 'acceptable' codes of behaviour in construction departments are linked to those of the construction industry. After all, the construction industry is where these students are eventually to belong if they are to succeed. It could be argued then that construction departments actively and/or passively promote and/or maintain the construction culture. If it is accepted that this culture is male and conflictual then it seems that the conflict will continue in future. If moves are made to change the cultures of the educational departments it could be argued that over time this may have an impact on the construction industry's culture. This could become particularly important as the proportion of the workforce that needs to be educated to a higher level increases, thus putting more importance on the role of construction departments.

What proportion of women in construction will create impact?

It is worth mentioning here that as the proportion of women in construction increases it does not follow that the culture will automatically change. The concept of critical mass; the proportion of a minority that will cause change to occur in the culture; customs; working practices and behaviour of a previously male-dominated situation is often argued to demand a proportion around 35 per cent (Kock, 1990). She develops her argument based on the theory expounded by Dahlerup (1988). The problem is that there is no empirical evidence in construction that this would be the case.

The theory states that when a 35 per cent proportionality is reached stereotyping will diminish, there will be new role models for girls and women, the open resistance towards engineer-women in the labour market will disappear, women's professional decisions will be trusted except be elderly people and female values will be accepted as appropriate and natural.

It has to be said that this theory is controversial and unproven. Just because women form an increasing proportion of the workforce it does not follow that female values will be promoted by them. One way in which women have learnt to cope in male-dominated organizations is to emulate male approaches.

The same can be said for men who would naturally prefer a different culture to the one they find themselves in. Because gender values can be described as a continuum ranging from male to female, it may be that men and women holding similar values are attracted to similar occupations. This would explain why not only women but men too may find the construction culture an unacceptable one in which to work.

It follows from this that the image of the construction industry may be an important factor in the career selection process of both young men and women. The image is based on the reality. If the reality is that the construction industry has a masculine culture then those who seek to be a part of that reality will join courses leading to careers in the industry. Further, there is then a vested interest in those who have chosen the culture to promote and maintain that culture and resist change. If male values include the propensity for conflict in human interaction, then conflict becomes locked into construction culture.

It seems that because of the likelihood of the male culture being perpetuated that construction culture may not be that much affected by an increase in the proportion of women over the next few years. Further, even a higher proportion of female construction graduates does not mean that they will end up in the most acutely male groupings in the industry because men and women alike seek their comfort zones within which to work.

If increasing the proportion of women in the construction industry cannot necessarily be relied upon to change the culture to one that is more female and thus less conflictual the question must be: How can the culture be changed?

The process of changing, in this case feminizing, the construction culture is a long-term prospect. The only real potential for meaningful change must surely be in the construction departments of the education system. One problem here is that the very nature of the vocational orientation of these departments presents a barrier to change. Courses vocationally orientated towards construction may well be seen in the long run to be counter-productive because by serving the industry's needs now future needs may remain unaddressed. This could arise because change would not be allowed to occur due to the vested interests of current construction culture.

183

This means that the vocational orientation of courses, or at least departments, may need to be questioned.

Conclusions

Women are under-represented in the construction industry in both contracting and professional offices. The cultures of the professions concerned vary but tend to be conservative and old-fashioned (Gale 1990).

The industrial and/or professional image plays a central part in career choice. However, if moves are made to change image then reality must also be changed otherwise the long-term effect of trying to attract more women into these male-dominated construction professions will not be achieved.

Female school students, parents, careers advisers and teachers remain sceptical of moves to attract women because many tend to view these initiatives as driven by purely economic motives which could be reversed at a later date.

It seems that if professions become more attractive to women then they may also attract more men. Whether this will mean over the long term that men will continue to dominate the construction industry is not clear. Patriarchal society is reflected in the organization of industry and contributes to the gendering of work. Male domination will be difficult to change. Organizations tend to reproduce themselves, maintaining male values. The construction industry culture is in need of change if it is to escape from its current characteristics of crisis, conflict and masculinity. The culture is characterized by male domination, aggression and conflict, gallant behaviour and traditional attitudes.

Education is a gatekeeper to the construction culture. The masculine culture is actively and passively promoted. Higher and further education provide a place where would-be construction managers learn to behave and identify with the dominant culture of the construction industry.

Initiatives designed to influence careers advisers and school students appear to work well in providing information, insights and images of the construction industry but may only reflect the reality of a few large firms. These initiatives may also act as a screen to ensure that more but 'appropriately orientated' students who identify with the construction culture apply for places at university to read construction degrees. However, CITB Construction Careers Service's recent careers publications and their commitment to encouraging more women into construction at craft technician and professional levels shows serious intent. Many others significantly affect the career choices of young men and women. These include parents, peers, careers teachers, careers advisers and the industry itself. Events and experiences in early childhood play an important part in influencing the later potential for a person to be changed or influenced. Therefore, although school students may make their career decisions at 15 years of age their fundamental predisposition is a function of early childhood.

Female construction undergraduates and graduates are almost invariably at pains to point out that they are not feminists and seek to distance themselves from feminists and feminism. They are fiercely independent. There is a contradiction in this position and the concerns female undergraduate and graduates express about career opportunities and the attitudes of male colleagues towards career breaks for women. Female construction students have benefited from the feminist struggle both in terms of their rights under the law and changes in working practices and expectations brought about by a constant challenge to male domination at work and in society.

Female undergraduates and graduates often appear to find it important to be 'one of the boys'. This produces a tension between their female identity and the male norms of the construction industry work culture. Women have to fit into the construction industry. Males may also not identify with the masculine construction culture. Both sexes have to 'fit in'. The results and conclusions of this study point to the strong possibility that men and women converge in the images they hold and their perceptions of the construction industry.

Educational departments of construction in further and higher education have masculine cultures and act as gatekeepers to the construction culture. This culture must feminize if a real change is to occur with respect to the problem of conflict in the industry. However, it seems that it is in the interests of those who have chosen to work in the industry to maintain the maleness of the culture, thus keeping conflict and crisis as preferred aspects of working life.

Attempts to limit the conflict endemic in the industry through alternative dispute procedures, however well meaning, are bound to have only a cosmetic effect because they are inherently superstructural. This means that such approaches and procedures are concerned with the periphery or symptoms and not the causes of conflict.

Male domination of construction is easily demonstrated statistically. Arguably the construction process is gendered as a male culture. This benefits patriarchal society but not the working process or product. There are social, philosophical, economic and psychological reasons to change the construction industry but this requires great vision and selflessness from many in the industry who may see that they stand to lose something if changes occur. In the future the construction industry is almost certain to come under close scrutiny from government as attitudes and expectations in society continue to expect equality in the work place.

Questions

1. Suggest measures which could be taken to improve recruitment and employment practices in order to attract and retain more women to the construction industry. Your answer should explain the importance of sex stereotyping and link this with practical approaches in personnel procedures with reference to relevant legislation.

2. 'Women continue to be significantly under-represented in the primary sector (agriculture and energy and water), in most manufacturing, in transport and communications and, in particular, in the construction industry.' Equal Opportunities Commission (1990).
   Discuss this statement and review theoretical arguments for why this situation prevails.

3. It has been argued that courses in construction and the departments that run them act as gatekeepers to the construction industry's culture. Explain what is

meant by this and suggest ways in which educationalists could play a proactive role in increasing the proportion of women in the construction industry.

References

Beacock, P. M. and Pearson, J. S. D. (1989) *Characteristics of higher education for the construction professions: development services project report 24*, London, CNAA, 96–7

Baker, E. and Gale, A. W. (1990) *Women in construction management: a report on two pilot insight programmes*, London, South Bank Polytechnic, 28 pages

Barron, R. and Norris, G. (1976) Sexual divisions and the labour market, in Barker, D. and Allen, S. (Eds) *Dependence and exploitation in work and marriage*

Carline, D. (1985) Trade unions and wages, in Carline D., Pissarides, C. A., Stanley Siebert, W. and Sloane, P. J. (Eds) *Labour economics*, Harlow, Longman, 186–224

Carter, R. and Kirkup, G. (1990) *Women in engineering, a good place to be?* London, Macmillan

Chiplin, B. and Sloane, P. J. (1976) *Sex Discrimination in the Labour Market*, London, Macmillan

CITB (1988) *Factors affecting the recruitment for the construction industry – the view of young people and their parents,* a research study prepared for the Construction Industry Training Board by Harris Research Centre, Bircham Newton, CITB

CITB (1993) *The Construction Industry Professional and Management Careers Handbook, 1993–1994.* East Grinstead, Construction Industry Training Board, 165–73

CIOB (1992) *Appendix A, Document AU(WIBCC)823*, Ascot, CIOB

Cockburn, C. (1983) *Brothers: Male domination and technical change*, London, Macmillan

Cockburn, C. (1985) *Machinery of Dominance: Women, Men and Technical Know-How*, London, Pluto Press, 167–89.

Cockburn, C. (1987) *Two track training, sex inequalities and the YTS*, London, Macmillan, 64–7

Cooper, C. L. and Davidson, M. (1984) *Women in Management*, London, Heinemann

Cuthbert, J. (1993) *Personal Communication*, Darlington, Department for Education

Dahlerup, D. (1988) *Vi har vent lange nok*, Kopenhaven

Dale, A. and Glover, J. (1990) *An analysis of women's employment patterns in the UK, France and the USA*, DoE Research Paper No. 75, London, HMSO, 15–16

Davidson, M. J. and Cooper, C. L. (Eds) (1984) *Working Women: an International Survey*, London, John Wiley & Sons

Dex, S. (1985) *The sexual division of work: conceptual revolutions in social science*, Brighton, Harvester Press

Dex, S. (1987) *Women's Occupational Mobility: a lifetime time perspective*, Basingstoke, Macmillan

Dex, S. (1988) *Women's Attitudes Towards Work*, London, Macmillan

EOC (1986) *Men's Jobs, Women's Jobs*, London, HMSO

EOC (1987a) *Equality Between Sexes in Industry; how far have we got?* London, HMSO

EOC (1987b) *Guidance on equal opportunities policies and practices in employment*, London, HMSO

EOC (1988) *Women and Men in Britain: a research profile*, London, HMSO, 37

EOC (1990) *Men and Women in Britain 1990*, London, HMSO, 13–16

Farish, M. (1987) *Personal Communication*

Garrett, S. (1987) *Gender*, London, Tavistock Publications, 1–28

Gale, A. W. (1987) The socio-economic significance of an increase in the proportion of women in the construction industry, in *Coombe Lodge Report, Managing Construction Education.* Vol. 20, Bristol, FESC, 113–24

Gale, A. W. (1989) Attracting women to construction, in *Chartered Builder.* Vol 1, No. 1, 8–13

Gale, A. W. (1991) What is good for women is good for men: theoretical foundations for action research aimed at increasing the proportion of women in construction management. In Barrett, P. and Males, R. (Eds) *Practice Management: New perspectives for the construction professional*, London, Chapman & Hall, 26–34.

Gale, A. W. (1992a) The construction industry's male culture must feminize if conflict is to be reduced: the role of education as a gatekeeper to a male construction industry, in Fenn, P. and Gameson, R. (Eds) *Construction Conflict: Management and Resolution*, London, F. N. Spon, 416–27

Gale, A. W. (1992b) Women in Construction: Reflections on findings and recommendations of two recent evaluation exercises on experimental insight courses for school students in Britain, in *Contributions to the Gender and Science and Technology Conference*, October 25–29, Eindhoven, Vol. 2, 335–44

Gale, A. W. (1993) *Women in Construction: An investigation into the under-representation of women in construction management in Britain*, PhD Thesis (in preparation), Bath, University of Bath

Greed, C. (1991) *Surveying sisters, women in a traditional male profession*, London, Routledge 51

Hakim, C. (1979) *Occupational segregation: A comparative study of the degree and pattern of the differentiation between men's and women's work in Britain, the United States and other countries*, Department of Employment Research Paper No. 9, London, HMSO

Handy, C. (1985) *Understanding Organizations*, Harmondsworth, Penguin

Hearn, J. (1990) The sexuality of organization, in Hearn, J., Sheppard, D. L., Tancred-Sheriff, P. and Burrell, G. (Eds) *The Sexuality of Organization*, London, Sage, 1–28

Holton, V. (1989) *The Female Resource – an Overview*, London, Women in Management

IPM (1987) *Contract compliance: The UK Experience*, an IPM/IDS report, Old Woking, Graham Press

Kock, H. (1990) *GASAT 1990 Conference Report Book*, Jonkoping, GASAT

Madden, J. F. (1973) *The Economics of Sex Discrimination*, Lexington, Mass, Lexington Books

Manley, P. and Sawbridge, D. (1980) Women at work, in *Lloyds Bank Review*, No. 135, 29–40

Martin, J. and Roberts, C. (1984) *Women and Employment: A Lifetime Perspective*, London, HMSO

Meehan, E. M. (1985) *Women's Rights at Work: Campaign and policy in Britain and the United States*, London, Macmillan

Oakley, A. (1972) *Sex, gender and society*, London, Gower

Oakley, A. (1981a) *From Here to Maternity: Becoming a mother*, Harmondsworth, Penguin

Oakley, A. (1981b) Interviewing women: a contradiction in terms, in Roberts (Ed) *Doing Feminist Research*, London, Routledge and Kegan Paul

Orr, P. (1985) Sex bias in schools: National perspectives, in Whyte, J., Rosemary, D., Kant, L. and Cruikshank, M. (Eds) *Girl-friendly Schooling*, 7–23

Pauli, R. (1989) Masculine management in industry, in *Harnessing the Female Resource; the women's education conference of 1989*, report of the conference held at the Queen Elizabeth II Conference Centre, July 11, London, Women In Management, 22–3

PCAS (1987) *Statistical Supplement of the PCAS Annual Report 1985–1986*, Mansfield Polytecnics Central Admissions System

PCAS (1988) *Statistical Supplement of the PCAS Annual Report 1986–1987*, Mansfield Polytecnics Central Admissions System

PCAS (1989) *Statistical Supplement of the PCAS Annual Report 1987–1988*, Mansfield Polytecnics Central Admissions System

PCAS (1990) *Statistical Supplement of the PCAS Annual Report 1988–1989*, Mansfield Polytecnics Central Admissions System

PCAS (1991) *Statistical Supplement of the PCAS Annual Report 1989–1990*, Mansfield Polytecnics Central Admissions System

PCAS (1992) *Statistical Supplement of the PCAS Annual Report 1990–1991*, Mansfield Polytecnics Central Admissions System

PCAS (1993) *Statistical Supplement of the PCAS Annual Report 1991–1992*, Mansfield Polytecnics Central Admissions System

Peacock, S. and Eaton, C. (1987) *Women in Engineering RC19*, Watford, EITB

Pearson, J. (1991) *Women in Construction*, Ascot, CIOB, DMX Thesis

Pyke, J. (1991) *Personal Communication*

Rainbird, H. (1989) *Personal Communication*

Rodgers, S. (1991) *Women in Construction*, BSc project Department of Civil Engineering, Queens University, Belfast

Sloane, P. J. (1985) Discrimination in the labour market, in Carline, D., Pissarides, C. A., Stanley, Siebert W. and Sloane, P. J. (Eds) *Labour Economics*, Harlow, Longman, 78–158

Spencer, A. and Podmore, D. (Eds) (1987) *In a Man's World; essays on women in male dominated professions*, London, Tavistock Publications

Spender, D. (1985) *For the record*, London, The Women's Press, 138

Srivastava, A. and Fryer, B. (1991) Widening access: women in construction, in *Proceedings of the Seventh Annual Conference, 1991*, September 9–10, 178–90

Srivastava, A. (1992) Gender and science and technology 1992, in *Contributions to the Gender and Science and Technology Conference*, October 25–29, **Vol. 1**, 175–84

Tsuchigane, R. and Dodge, N. (1974) *Economic Discrimination Against Women in the United States: measures and changes*, Lexington, Mass, Lexington Books, 5

UFC (1990) *Universities Statistical Record 1988–89: volume 1, students and staff*, London, Universities Funding Council

UFC (1991) *Universities Statistical Record 1989–90: volume 1, students and staff*, London, Universities Funding Council

UFC (1992) *Universities Statistical Record 1990–91: volume 1, students and staff*, London, Universities Funding Council

UGC (1987) *Universities Statistical Record 1985–86: volume 1, students and staff*, London, Universities Grants Committee

UGC (1988) *Universities Statistical Record 1986–87: volume 1, students and staff*, London, Universities Grants Committee

UGC (1989) *Universities Statistical Record 1987–88: volume 1, students and staff*, London, Universities Grants Committee

Walby, S. (1986) *Patriarchy at Work*, Oxford, Polity Press, 247

Walby, S. (Ed) (1988) *Gender Segregation at Work*, Milton Keynes, Open University Press

Walker, D. J. (1975) *Sex Discrimination: A simple guide to the complicated provision of the Sex Discrimination Act*, London, Shaw & Son Ltd.

Wells, J. (1990) *Female Participation in the Construction Industry*, Geneva, International Labour Office, 28

West, L. J. (1982) *An investigation into the career choice and occupational perspectives of professional builders and quantity surveyors*, PhD, Heriot-Watt University, 1–10

WICAG (undated) *Recruiting and Employing Women; a guide for construction employers*, London, Women In Construction Advisory Group

Wilkinson, S. (1992a) Giving girls a taste of technology, in *Contributions to the Gender and Science and Technology Conference*, October 25–29, **Vol. 2**, 383–94

Wilkinson, S. (1992b) Career paths and child care: employer's attitudes towards women in construction, in *Proceedings from the Women In Construction Conference* at the University of Northumbria, September 8

WAMT (1993) *W.A.M.T. Newsletter*, Spring, London, Women & Manual Trades

Chapter 10

Directions in human resources

This chapter is speculative in nature. It attempts to suggest an agenda for the future of HRM in the construction industry. The scene is briefly set with reference to demographic trends. Structural changes in the construction industry are linked to the economic environment and the recent reduction in training.

The future of the built environment and demand for construction is discussed in relation to changes in the nature and pattern of work. Recent research on organizational change is critically reviewed and the question of the paternalism and image of the construction industry are discussed with respect to recruitment and productivity.

Telecommuting as a potential new working style is discussed in relation to the process and future of the construction industry.

Education, training and continuing professional development receive attention. Reference is made to recent reports on the importance of moving towards a common education for built environment professionals and the significance of National Vocational Qualifications (NVQ).

Setting the scene

Kivisto (1987) describes the three-wave theory of history based on Toffler's discourse (1981) shown in Figure 10.1. He argues that the first wave (agriculture) began around 10,000 BC as permanent settlements sprang up. Work and organization were related to agricultural production, storage and distribution throughout this wave. The eighteenth century saw the birth of the industrial revolution – the second wave of history. Science and technology became fundamentally important during this wave. The opportunities offered by new machines enabled entrepreneurs to develop industrial products and production. Marx would argue that capitalist production is primarily concerned with the production of commodities which have the dual function of use and exchangeability (Marx 1867). The third wave (service and information) began recently. It follows the dominant need for human interaction and communication on a global scale.

Kivisto (1987) states that construction is always part of human activity and culture. In the earliest period of agricultural society it was necessary to provide shelter for animals and people. Kivisto says that the future can be forecast on the basis of reflections on the past and consideration of the present. Toffler (1981) says that

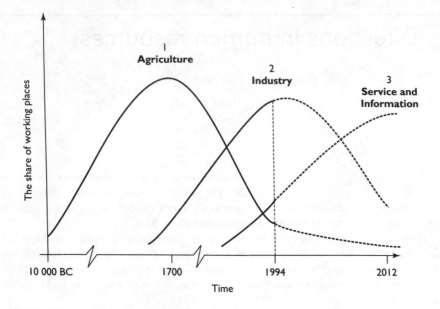

Fig 10.1 The three waves of history. (*Adapted from* Kivisto (1987) and Toffler (1980))

where more than one wave is 'breaking' there is conflict, contradiction and confusion but through the analysis of the political and social forces acting a clearer image of the future is possible. At this time in western history there are three waves in action. In some developing countries like Korea there may be a jump directly from an agricultural to an information society. All of this has a bearing on construction HRM.

Kivisto (1987) maintains that in western societies future trends in construction will be influenced by patterns of urbanization. One scenario is that of a highly urbanized society. Handy (1985) argues that in this type of society information would be all important and people would live in a closed world. There would be an elite of highly-paid, well-educated professional people. Skilled labour and middle management jobs would have very much reduced and a grim segregated society would develop based on a free market free-for-all with restricted travel between regional zones; even in small countries like Britain.

The alternative scenario put forward by Kivisto (1987) is that of higher employment with people working mainly from home using information technology (IT) to communicate. People would live outside cities in rural areas, revived by the buying power of the new rural knowledge workers. This would coincide with increased life spans, a reduction in child birth and an increase in the number of households. He predicts contradictions between the needs, values and attitudes of different population groups. These factors will affect the nature and volume of construction. Also there is predicted to be a major shift in demand (already begun) for construction to maintenance. A move away from new buildings to rebuilding and the modification of existing building stock.

The economic, social and sanitary, etc. significance of the existing structure will increase. For instance, the spreading of city structures results in time in increased operational costs curtailing, e.g. the communes' chances to invest in new structures. (Kivisto 1987.)

In so-called developing countries there is predicted to be a transition from agricultural to industrial society; an increase in population size and increased urbanization. Demand for construction will grow massively but will not be met because of a short fall in resources; especially due to an under-educated human resource.

Demographic factors and change

Between 1987 and 1995 NEDO (1988) predict a fall of 20 per cent in 16–24 year olds with 16–19 year olds falling 23 per cent – a function of a drop in the birth rate during the 1970s. Competition by employers for school leavers, marked by the 1988–1993 recession will intensify. At the time of a rise in the need for an increase in education and training there is a reduction in young people. Total construction output in the UK was predicted (Dove 1993) to fall by 6.9 per cent from 1991 to 1994 based on government figures (£31,884m in 1991 to £29,670m in 1994, with a low of £29,295m in 1993 (1985 prices)).

The number employed in construction is notoriously difficult to quantify, Ball (1988) states that in 1973 there were 2 million unemployed construction workers and in 1984 the figure had fallen to 1.5 million with an estimated 500,000 unemployed building workers. In 1991 and 1992, 813,000 and 910,000 respectively were employed in construction (Dove 1993). It would seem likely then that there are roughly 1.2 million unemployed building workers in the UK.

Ball (1988) says that the skills structure of the construction industry is riddled with status and skills divisions. He cites the Construction Industry Training Board (CITB) as saying that 70 per cent of construction workers in private sector employment are skilled workers. Skill can be defined as follows:

A social construct denoting status, earning capacity, industrial power and the ability to exclude others, as well as indicating a capacity to undertake certain specific tasks. (Ball 1988.)

CSSC (1988) reports that the structure of the construction industry and its employment characteristics are changing. The need for contractors to reduce overhead costs has lead to an increase in the use of subcontractors and self-employed labour. The number of construction workers holding 714 certificates rose from 150,000 in 1977 to over 400,000 in 1987. The number of trade apprentices has fallen sharply. The larger firms train and the smaller firms avoid the costs of training, thus enabling them to undercut those firms whose overheads contain a significant component of training costs. There is a structural change occurring in the industry. The number of small firms (8 or less employees) rose by over 140 per cent from 64,228 in 1976 to 155,624 in 1986. Large firms declined by over 42 per cent in the same period

191

(182 in 1976 to 105 in 1986). A high proportion of these small firms are specialist subcontractors, particularly in building services engineering. Buildings are becoming more complex and have an increasing proportion of building engineering services (Kivisto 1987).

Training is crucial but the picture looks problematic. This trend towards the emergence of small companies is referred to by Handy (1985, 1989). He argues that firms will in future concentrate on their core business and cites the construction industry as a good example of the structural trend towards subcontracting.

Productivity is very low in the UK. According to Andrews and Derbyshire (1993) the British construction industry has a lower productivity than the US, Japan, France and Italy. Britain has the lowest earnings out of these countries and building costs for industrial and office construction are higher in Britain than any other developed country except Japan. They link low pay with poor performance.

The need to plan ahead

NEDO (1988) argues in a recent report that firms should plan ahead. Competition between employers for young people will be fierce in the future. There may be some successful firms but there will also be many unsuccessful ones. The report suggests strategies for change in order to adapt to the effects of future trends. However, it is pointed out that many of these require a long lead-time. Therefore action is required urgently to prevent future disasters.

Actions include the need to increase the attractiveness of career prospects to young people. CITB (1988) also make this point. It is going to be difficult for the construction industry to achieve this as its image reflects an unsatisfactory reality; a view shared by the National Contractors' Group (NCG), according to Pettipher (1992). He cites the *Building 2001* report (CSSC 1988): 'Few people or organizations take the issue of image sufficiently seriously because no direct correlation between image and profitability is seen to exist'. He also cites the view of consultant corporate image-maker Graham Whitehead of Peter Prowse who believes that the construction industry faces an uphill battle with its image due to its diversity leaving it less susceptible to slick presentation compared to industries like retailing. Pettipher quotes Terry Tyrell, chair of image makers Sampson Tyrell: 'Construction has a slapdash image because people don't think of the industry collectively. They view it in terms of the one or two companies they have experience of and if that's a bad experience it tarnishes the whole industry.'

NEDO (1988) suggests that recruitment criteria should be modified in order to recruit those who would previously have been excluded. Training can then be used to rectify any short falls in experience and knowledge.

Existing staff can be retrained. This particularly applies to women who are almost invariably employed in traditionally 'women's jobs' (secretarial and clerical). They often have highly transferable skills and may in many cases be graduates. Their knowledge of the industry process and practices has been assimilated whilst working in construction. They represent a very good retrainable resource. Increased retention rates are seen by NEDO (1988) as a viable way of stabilizing the labour force. The three Rs of recruitment, reward and retention must all be focused upon.

Equal opportunities for under-represented workers

NEDO says that firms should widen their recruitment catchment geographically. Perhaps more important is the question of improving recruitment through imaginative equal opportunities policies and their implementation. Chapter 9 highlights the low proportion of women in the construction industry. People from ethnic minorities are also under-represented. The Commission for Racial Equality (CRE) has strongly criticized the construction industry for actively discriminating on racial grounds (Alexander 1992). The CRE says that the construction industry has made 'limited progress in implementing equal opportunities policies' with little evidence of ethnic monitoring or positive action. Some housing associations are insisting on the contractors they employ monitoring their subcontractors' ethnic make-up as well as their own if they are to remain on approved contractor lists. The insistence by clients on the provision and implementation of equal opportunities policies by contractors will probably increase in the future. The clients will not accept window-dressing and will demand evidence of positive action. The white male image of the construction industry referred to by the CITB report in 1988 will come under increasing scrutiny in the future. The construction industry and society as a whole will benefit from the outcome of this scrutiny as employment practices are forced to improve, albeit due to pressure from clients and politicians.

NEDO (1988) also suggests recruiting from sources of labour other than young people. Adults should be recruited particularly the long-term unemployed, returners (e.g. women who have taken time out of the labour market to rear children), former employees and retired workers.

A lot of the suggestions put forward by NEDO are concerned with elements of the labour market which have negative stereotypes. For example, people who are classified are long-term unemployed, women and people from ethnic minorities. For employment policies in the construction industry to change there will need to be a real change in construction industry culture. This will only be brought about by the combined effects of public opinion, political and client pressure.

The future working life

Handy (1985 and 1989) says that in the past we used to start our working lives with an employer with which we assumed life-long employment. We would sign on for what he calls a 100,000-hour job; (47 years ×47 weeks per year ×47 hours per week = 103,823 hours, approximately 100,000). He argues that in order that everyone has work in the future people will need to take on the notion of the 50,000 hour job (half the hours of a traditional job). This new type of job will be fragmented. It will start later in life after a longer education and training. People will work for several employers and make a career change two or three times in a working life. There will be a need to retrain and to take a large proportion of time in being educated and training whilst working – about 20 per cent of working time. Finally people will retire earlier.

The apparent contradiction with what has been said before about skills shortages is explained through understanding that the new work content will be highly skilled but the numbers employed to carry it out will be much less. Therefore there will be a smaller amount of work to go round but that work will require highly trained people. The new 50,000 hour jobs will be for 37 years, for 37 weeks of the year and for 37 hours per week. ($37 \times 37 \times 37 = 50,653$ hours, approximately 50,000.)

This new work pattern does not sit well with the construction industry tradition which thrives on very long and unsocial hours. Typically, the site hours are 10 hours per day and frequently involve Saturday working. The level of uncertainty in construction activity durations means that high levels of overtime are often necessary in order to meet programmes and deadlines. Systematic overtime is a feature of the construction industry. On overseas projects it is not uncommon for the normal working day to be 10 or even 12 hours in a 6 or even 7 day week. Arguably people adapt their productivity to meet the demands of such long working hours which are frequently worked to justify the levels of pay associated with onerous overseas conditions of contract. The working conditions and hours of construction projects are a very important part of the mythology of construction.

Handy (1985) argues that the informal economy, made up of the so-called 'black' economy (unaccounted illegal work), voluntary work and home-based work including DIY, will grow rapidly in the future. He estimates that this part of the economy already amounts to 40 per cent of the formal economy. It is with these trends in mind that the construction industry needs to plan future developments in HRM. Demand for construction work will be affected by these trends. The buying power and behaviour of people will determine demand. Also, the labour market will be affected which will in turn influence the productivity and profitability of the construction industry.

Handy (1989) criticizes the policy, adopted by many firms, of 'homegrown' workers and managers. Large, paternalistic construction firms are keen on this approach for training both school leavers and graduates. Handy argues that firms recruiting 16-year-olds and training them in the company will end up with staff at 30 with what he calls 'a severe case of group-think – well-meaning people too close and cohesive to challenge assumptions'. This runs counter to the notion of new work with fragmented career patterns.

Organizations of the future

The future of work is inextricably linked to the future of organizations. Kvande and Rasmussen (1992) say that there are two 'ideal' types of organization:

1. static hierarchies
2. dynamic networks

Static hierarchies are an old-fashioned system in which older men maintain patriarchal power relations. Young men are controlled by old men. The construction industry would appear to fit this model.

The dynamic network organization is one in which the dominant male group are young men who form alliances with 'new' women, who are 'like themselves' in work orientation and motivation. These young alliances usurp the older men. Women get on better in these organizations compared with static hierarchies.

Construction organizations tend to develop along traditional military hierarchical lines with a dominant older male control system. If more women are to be attracted to construction (CSSC 1988, NEDO 1988, Handy 1985 and 1989) the newer dynamic network organizational form will need to develop in the construction industry. If construction appears unattractive and inflexible to potential recruits by not having dynamic network-type organizations it will face problems in recruiting people who may prefer the opportunities and working environment offered in other industry sectors. Kvande and Rasmussen state:

> In dynamic networks the work is organized in teams where all members are equally important and all contribute knowledge and effort on an equal basis. As they work, the graduate engineers get to know each other's academic and personal qualifications. The women become visible as professionals for their colleagues and superiors.

Kvande and Rasmussen cite Kanter (1989):

> Companies are moving away from diversification towards maximizing their core business competence. They develop delayering the hierarchy and making the company leaner.

Handy (1989) addresses this subject in depth. He says that the new organization will be like a shamrock with three leaves. There will be a central core in which essential highly-paid professionals will organize the strategy and work of the core and the other two leaves. These other two leaves will consist of specialist subcontractors performing manufacturing and assembly functions and a flexible temporary, often part-time, workforce. He also states that the guiding principles of the new organizations will be intelligence, information and ideas. These are not the philosophies of the old factory with its military-style organization (see Pauli 1989, Chapter 9).

The construction industry will be affected by these organizational trends. There is already some evidence of a reduction in the number of large firms and an increase in the number of small ones. Forms of building procurement system are changing and although the so-called management contracting procurement system, described by Ball (1988) as 'the ultimate extension of subcontracting to all work', has been promoted by contractors it is not as yet as popular as traditional and design-and-build procurement systems (Masterman 1992).

Construction firms are paternalistic and hierarchical. The influence of operational management styles employed in construction project management has a strong influence on the industry's management culture. The future of construction organizations will almost certainly be one in which traditional organizational values and forms become increasingly challenged by a changing society.

Telecommuting – threat or opportunity?

Guest (1990) defines the style of working called telecommuting as relying on 'relatively cheap and simple technology – a personal computer a modem, a unit that connects computers by telephone'.

Handy (1989) cites examples of telecommuting organizations such as F International, set up in 1962, with a turnover in 1988 of nearly £20 million and 1100 employees of which 705 worked from home or local work centres. This company said that their employees' performance was over 30 per cent higher than their counterparts working in offices. Since this time the company has shed workers, almost certainly due to the recession (1988–1993). Eminent company presidents in the US have stated that between 25 and 75 per cent of their work could be done from home using appropriate communications (Handy 1989 and Toffler 1981).

Guest (1990) suggests that by the year 2000 half of the people working in London will spend part of their week working at home. Currently 40 per cent of white collar workers regularly use computers. It is envisaged that London will be the centre for this new telecommuting trend because of the high cost and stress of physically commuting.

Guest (1990) says that employers acknowledge that work patterns are changing and that firms need to introduce more flexibility into their structures. He disputes the notion that humans are social animals, citing as evidence single occupancy commuter cars and rapid growth in the use of personal stereos. Also he says that the quality of our lives will be improved by the removal of unpleasant and stressful commuting at the beginning and end of the working day.

Handy (1989) suggests that telecommuting offers increased flexibility for both the organization and worker. However, he acknowledges that some may not like this style. He cites examples of unhappy lonely telecommuters. If this is considered in the context of homeworking, which Handy himself identifies as the historical beginning of this style of working, there are potential disadvantages of isolation, fragmentation and loss of rights as far as workers are concerned. A superficial analysis of homeworking or telecommuting may give the impression that it offers access to work opportunities for women. However, this form of homeworking was seen as potentially marginalizing to women in paid work. It served as a device to keep them in the home and to continue to be able to have a dual career: looking after children (unpaid informal economic activity) as well as formal paid employment.

There are arguments in favour of regular social contact. This would be sharply reduced by the widespread introduction of telecommuting. Peters and Waterman (1982) argue in favour of a robust and powerful approach towards the process of communication. They say, using IBM as a case study, that facilities in offices, such as blackboards to aid communication, help promote informal interaction, innovation and change. Common dining areas also play an important part in providing the furniture for social interaction.

Allen (1979) shows from his research that there is a direct relationship between separation distance and the probability of communication at least once a week. This is based on relatively short distances (up to 100 metres). However, if we extrapolate the trend it would appear that much greater distances, an inevitable feature of telecommuting, would reinforce this lack of regular informal communication. It could be argued that this would in turn lead to a reduction in the level of innovation.

Implications for construction

It would appear that corporate headquarters jobs are more likely to be affected by a move towards telecommuting. Recruitment to the construction industry will be affected by the approach taken by the industry to telecommuting. The expectations of would-be recruits in relation to working styles will probably have an influence on the implementation of information technology and the extent to which telecommuting is adopted. Also the potential for reducing overheads through a high proportion of staff working from home will be considered by the corporate accountants. It is difficult to say which way things will go as there may be a general desire in the workforce for association and social contact at work. Also, the economics of innovation could find against telecommuting. The nature of site working makes it difficult to see how telecommuting could denude the construction site of staff. Gale (1991) argues that if the construction industry cannot compete or demonstrate specific advantages over the working styles of other industries recruitment to construction may become more difficult.

The higher profit margins of British contractors compared with those of other European countries may force the economic advantages of telecommuting for head office employees on to the agenda. Costs at the end of the day will be the driving force for change in the construction industry. Although tending towards the conservative and inflexible paternalism of the construction industry may well be the basis for resistance to changes involving new information technology. Koontz *et al.* (1984) state that psychological inflexibility in 'managers and employees in organizations may develop patterns of thought and behaviour that are hard to change'.

It might be argued though that the construction industry has a good and developed tradition of communicating with sites remote from regional and national offices. This is a central characteristic of the construction industry. In this sense no real problems could be envisaged if this highly-developed expertise in distance communication may be extended into the corporate arena.

There seems to be a reasonable case for telecommuting in the construction industry to be the subject of research in order to establish likely future directions for organizational change, working style and recruitment (Gale 1991).

Education and training

J. M. K. Laing, chair of the National Contractors Group (NCG), wrote in the foreword to *Building 2001* (1988) that there were six major challenges facing the construction industry. The sixth is concerned with falling numbers entering the UK labour market and changes in training methods and employment.

As already mentioned, trade apprenticeships have been declining in number for over ten years now. Just prior to the recession (1988–1993) skills shortages were acute, especially in the south-east. Training has reduced because of the combined effect of a reduction in work load coupled with an increase in self-employment (CSSC 1988). Increased competition and lump sum contracts have contributed to an unwillingness to train due to costs and the perceived and real poaching of trained, skilled and professional workers from firms that train by firms that do not.

Gale and Fellows (1990), reporting on a conference organized by the Association of Researchers in Construction Management cite a previous minister of Higher Education, Mr Robert Jackson (CIC 1989), who said 'Traditional

197

self-contained qualifying courses in the construction industry are no longer adequate.'

Human beings are the most important resource we have because they are the source of innovation. Dr Elizabeth Nelson stated in concluding remarks of her recent Royal Society of Arts lecture that we have to be ready to take the challenge of the new society. She epitomized this as displaying the growth in trends of pleasure, complexity, networking, open citizenship, a greater self-knowledge and control of strategic opportunism and the exploration of new mental frontiers.

Demographic, social, political and technological changes dictate the need to educate, update, develop and retain skilled personnel. There is a need to improve co-operation and collaboration between industry and higher education and professional institutions.

Professional education and training

Sir Karl Popper, according to Burke (1983), the most influential philosopher of science this century, said:

> We are not students of some subject matter but students of problems. And problems may cut right across the borders of any subject matter or discipline (Popper 1974).

This view can be usefully applied to the problem of producing, reproducing and maintaining our built environment. The number of professions, chartered and otherwise, is too many according to many influential voices among clients, government, academia and the construction industry. For some time there has been a growing debate about the need to develop a common educational programme for all built environment professionals. The sluggishness of real change is probably due to the vested interests of the professions and some in higher education.

Government forced the construction industry to come together in order to speak with one voice under the umbrella Construction Industry Council (CIC), originally the Building Industry Council (BIC). It was formed in 1988 with the two main aims:

1. To strengthen the research base of the construction industry.
2. To improve professional education and training.

If the CIC had not been formed government would have imposed its own solution. The CIC now represents 24 professional bodies and associations and has five associate members:

British Association of Construction Heads
Construction Industry Training Board
Faculty of Building
Institute of Concrete Technology
Standing Conference of Heads of Schools of Architecture

The CIC published a report in 1993 (Andrews and Derbyshire 1993) which addressed the level of commonality and the potential for commonality in the education and training for construction industry professionals. The debate is hotting up. Andrew Ramsay, secretary of the Chartered Institution of Building Services Engineers (CIBSE), writing in *Building Services* (1993), argues for its members to join forces with building surveyors (Royal Institution of Chartered Surveyors, RICS, members) in order to challenge the traditional supremacy of the architectural profession, whose status as architects has up until now been protected by law under the Architects Registration Council for the UK (ARCUK). He speaks optimistically of the potential for government to support the abolition of ARCUK in the near future. This is evidence of inter-profession warfare.

Collier *et al.* (1991) were commissioned by the Council for National Academic Awards (CNAA) to look into interdisciplinary aspects of built environment curricula. The scope of their study included the fields of architecture, building, building services engineering, building surveying, civil engineering, estate management, housing, landscape architecture, quantity surveying, structural engineering and planning. They were asked to identify commonality in higher education courses, project work, potential for collaboration between disciplines and constraints. Their conclusion point firmly in the direction of a need for change to a common education for all built environment professionals. They state that academic institutions should develop a common culture for all built environment students and allow flexibility in access to and between courses of study.

Particularly important is their call for enhanced opportunities for research and postgraduate study and the provision of continuing professional development (CPD) courses and seminars in order to develop the skills and knowledge of built environment professionals throughout their working lives.

They state that the professional bodies should integrate and that courses should give multiple exemptions from the examinations required for membership of several institutions. Broadly speaking a call is made for unification not differentiation and the case is made for later specialization allowing for movement between several professional bodies. They say that the construction industry should collaborate with higher education and the professions.

Professor Hans Haenlein (1989), chair of the BIC in 1989, said *inter alia* defining the purpose of a meeting of course heads sponsored by the BIC:

> Concern has been expressed as to whether in the face of demographic decline the industry can attract and train sufficient people to meet the demands of the 1990s. The focus of all institutions within the industry must be to address the challenge of developing a framework of coherent vocational qualifications within which can be set the recruitment, education and training of the industry's labour force.

Haenlein (an architectural head and practitioner at that time) links the problems of recruitment, vocational education and training, the role of institutions and demographic trends together with the need for concerted action.

Andrews and Derbyshire (1993) were commissioned by the CIC to 'establish the scope for greater commonality and purpose in the education, training and continuing professional development (CPD) of construction professions.' They reported that fragmentation of the construction industry professions is seen from two conflicting perspectives. The majority view is that fragmentation of professional roles is a problem. The minority view is that fragmentation of professional roles is the inevitable necessity of specialization.

Andrews and Derbyshire conclude that there is scope for commonality in education, training and CPD.

National Vocational Qualifications (NVQ)

There are five levels of NVQ. 1, 2 and 3 are for craft and technician grades and the CITB has the role of the industry 'lead body' for these. They have been developed and are in use, receiving growing acknowledgment and acceptance. The Construction Industry Standing Conference (CISC) is the industry lead body responsible for the development of NVQs level 4 and 5 for professional work. The first phase of occupational mapping has been completed and the second phase involving the generation feedback on draft occupational standards for technical, managerial and professional roles was ongoing during 1993 (CISC 1993). Andrews and Derbyshire quote some of the responses they received with respect to NVQs from some of the 44 interviews they conducted with representatives from the construction industry professions, higher education and practice.

Some say that they see NVQs as an important tool in their own right and state that they can be used as a means of identifying greater commonality. However, in higher education and professional institutions their appropriateness from a top down perspective is challenged. There are many who feel that the case for NVQs at the professional level has yet to be made. There is an important difference perceived among many in higher education and the professions between operational levels of work covered by NVQs 1, 2 and 3 and professional levels of work covered by NVQs 4 and 5 involving judgemental decisions, long-term perspectives, intuition and creativity. Some might characterize this argument by saying that NVQs are inherently competency-based and therefore less appropriate to work required at professional level. Although knowledge and understanding are said not to be part of vocational standards they are nevertheless essential components at every level, particularly at the higher professional levels. Some say: 'How are they (vocational standards) to be specified, provided and assessed? How are the proper proportions of vocational and academic content to be decided – and by whom?'

Andrews and Derbyshire (1993) conclude:

> There is evidence that NVQs may assist in the long term but there is also a high degree of uncertainty in the industry about the ultimate practicalities of implementation, especially at professional levels.

Postgraduate education and continuing professional development (CPD)

Chartered and non-chartered professional institutions are actively promoting CPD already, monitoring their membership in various ways. According to Andrews and Derbyshire (1993) all the main chartered institutions, except the Institution of Civil Engineers who only recommend it, make CPD obligatory. Minimum requirements for CPD to be undertaken are between 20 and 60 hours per year, monitored by making it a requirement for the upgrading of membership.

A unifying theme for the built environment professions is CPD. The CIOB published a report on education (1989) which encourages the expansion of CPD. The two aims of the report were to increase the educational influence of the CIOB and to strengthen the discipline of building. The general tenor of the report could be interpreted as being away from the notion of commonality. However, the general trend towards commonality seems strong if slow.

Postgraduate education for construction professionals is a vital area for expansion in the UK. In many European countries, notably France, a first degree qualifies the holder for a technician role, such as a draftsperson. In the UK health professions there are parallels. In clinical psychology, for example, a taught masters degree (now three years in duration) is mandatory for professional practice and chartered status. In UK industry it is said that professionals can get nowhere in chemical engineering and pharmaceutical science without a doctorate. Compare this with advertisements that appear in the technical and trade press for senior civil engineers and project managers, requiring a good degree and or chartered status. These are often the highest requirements. Experience combined with a first degree or Higher National Diploma are also very common requirements for high level executive positions in the construction industry.

With growing competition in an internationalized construction market and the trend towards more complex buildings, and social and political pressures affecting the future labour market, the construction professional of tomorrow will need to be more of a polymath. This will not be achieved without significant changes to the education and training of built environment professionals.

Conclusions

The future of construction HRM is not easy to predict. However, issues relating to the future can be identified. One problem is that the construction industry is sensitive to the underlying economic cycle. This tends to change the agenda for senior managers in the construction industry from that of being able to look ahead during a time of boom to trying to survive in the depths of a recession.

The construction industry has forced government to retain the CITB and ECITB. This has enabled training to be far more organized and stable than it would otherwise have been had these Industry Training Boards been abolished like all the rest. Indeed the role of the CITB through its Construction Careers Service (CCS) has been nationally effective on the professional recruitment front. The construction industry has acted in concert through an innovative move taken by the National Contractors Group (NCG) to fund the provision of degree courses in a number of universities. The contractors concerned, not all of them that large, sponsor students to attend these courses as well as funding the universities directly. In the future in a time of relative boom the skills shortages and subsequent problems of quality and high levels of professional pay should be alleviated.

The future of work and organizations are important considerations for HRM in the future. The future expectations of school leavers, the working population as a whole and pressure from government will strongly influence trends in recruitment. Race and sex equality will become far more important for construction managers of the future to address seriously and effectively. Public opinion, government and clients will demand that under-representation of ethnic minorities and women in the construction industry be dealt with in a positive way.

Education and training will become more and more important and costly to the construction industry. The working relationship between the construction industry, higher and further education and professional bodies will need to develop to enable the integration and commonality demanded by clients and government to be achieved. Continuing professional development (CPD) and postgraduate education will grow in importance and demand a significant amount of time in the working lives of the construction professionals of the future.

The costs of all these changes, particularly education and training, will be avoided by some firms who will use the competitive advantage this gives to undercut firms who do train. It seems likely that government may be called upon to legislate for a level playing field, to regulate for fair competition.

The construction industry is an important industry now but its importance will grow as a major employer in a future society where work becomes increasingly scarce, more highly professionalized and fragmented.

Questions

1. Evaluate the wisdom and significance of increasing the commonality of education and training for built environment professionals. Link this evaluation to possible changes in the nature of work which are likely to influence recruitment of professional workers to the construction industry in the future.
2. Discuss the relationship between demographic trends, the structure of the construction industry, education and training. Make predictions of the likely effect on the construction industry of future requirements by clients, government and public opinion. Particular attention should be given to personnel policies and recruitment strategies.
3. Define the working style known as telecommuting. Develop an argument explaining how and why this style of working may be adopted in the construction industry over the next decade. Relate your argument to trends in the changing nature of organizations.

References

Alexander, D. (1992) HAs set to get tough on racial equality, in *New Builder*, October 22, No. 149, 3

Allen, T. (1979) Communication in the laboratory in *Technology Review*, October–November

Andrews, J. and Derbyshire, A. (1993) *Crossing Boundaries: a report on the state of commonality in education and training for the construction professions*, London, Construction Industry Council

Ball, M. (1988) *Rebuilding Construction: Economic Change in the British construction industry*, London, Routledge

BIC (1989) Educational futures for the construction industry, in *Proceedings of the First Heads of Courses Meeting*, May 24, London, Building Industry Council

Burke, T. E. (1983) *The philosophy of Popper*, Manchester, Manchester University Press

CIOB (1989) *Building Education for Tomorrow: Report of the education working party*, Ascot, Chartered Institute of Building

CISC (1992) *Occupational Standards for Technical, Managerial and Professional Roles in the Construction Industry (Consultation Edition May 1992)*, London, Construction Industry Standing Conference

CITB (1988) *Factors Affecting the Recruitment for the Construction Industry: the view of young people and their parents*. A study prepared for the CITB by Harris Research Centre, Bircham Newton, Construction Industry Training Board

Collier, A., Bacon, J., Burns, D. and Muir, T. (1991) *Interdisciplinary Studies in the Built Environment: a CNAA research project supported by the Department of Environment*, London, Council for National Academic Awards

CSSC (1988) *Building Britain 2001*, University of Reading, Centre for Strategic Studies in Construction

Dove, B. (1993) Economic report, in *New Builder*, May 14, No. 174, 20

Gale, A. W. (1990) Telecommuting, in *Proceedings of the Fifth Annual ARCOM Conference*, Salford, ARCOM, 95–100

Gale, A. W. and Fellows, R. F. (1990) Challenge and innovation: the challenge to the construction industry, in *Construction Management and Economics*, London, E. & F. N. Spon Ltd, **8,** 431–6

Guest, D. (1990) Telecommute and beat the rush-hour, in *The Sunday Correspondent*, July 15, 46

Handy, C. (1985) *The Future of Work*, Oxford, Blackwell

Handy, C. (1989) *The Age of Unreason*, London, Hutchinson, 84–7, 120, 155, 188–9, 190

Kanter, R. M. (1989) *When Giants Learn to Dance*, London, Simon & Schuster

Kivisto, T. (1987) Futures research forecasting in construction, in *Kenotes, CIB proceedings of the fourth international symposium on building economics*, Copenhagen, CIB W55/W82, 144–77

Koontz, H., O'Donnell, C. and Weihrich, H. (1984) *Management*, Singapore, McGraw-Hill, 216–7

Kvande, E. and Rasmussen, B. (1992) Structures – politics – cultures: understanding the gendering of organizations, in *Proceedings of GASAT East and West European Conference*, Eindhoven, October 25–29, GASAT, 93–128

Marx, K. (1867) *Capital Vol 1*, New York, International Publishers

Masterman, J. W. E. (1992) *An Introduction to Procurement Systems*, London, E. & F. N. Spon

NEDO (1988) *Young People in the Labour Market: a challenge for the 1990s*, London, NEDO

Peters, T. J. and Waterman, R. H. (1982) *In Search of Excellence: lessons from America's best run companies*, New York, Harper & Row, 219–20

Pettipher, M. (1992) Not looking good, in *New Builder*, October 8, No. 147, 16

Popper, K. (1974) *Conjectures and Refutations*, London, Routledge and Kegan Paul, 66

Ramsay, A. (1993) Keeping hold of nurse, in *Building Services*, May 13

Toffler, A. (1981) *The Third Wave*, London, Pan Books

Williams, C. (1990) We should be talking to each other!, in *Construction Computing*, Spring No. 29, 15, 20

Index